Astronomers' Universe

For further volumes:
http://www.springer.com/series/6960

Duncan Lunan

The Stones and the Stars

Building Scotland's Newest Megalith

 Springer

Duncan Lunan
Troon, Ayrshire KA10 6JE

ISSN 1614-659X
ISBN 978-1-4614-5353-6   ISBN 978-1-4614-5354-3 (eBook)
DOI 10.1007/978-1-4614-5354-3
Springer New York Heidelberg Dordrecht London

Library of Congress Control Number: 2012947404

Printed on acid-free paper

Springer is part of Springer Science+Business Media (www.springer.com)

To the late Professor Alexander Thom,
With deepest respect,
Glasgow Parks Department Astronomy
Project.
"Hats off lads, we're beat."

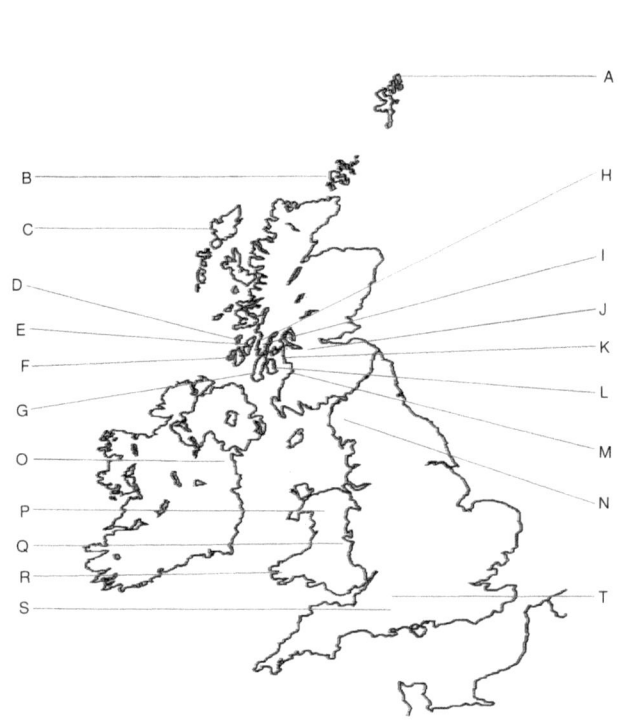

A

B

C

D

E

F

G

H

I

J

K

L

M

N

O

P

Q

R

S

T

Key to Map of British Isles

A. Shetland (Unst).

B. Orkney (Ring of Brodgar, Ness of Brodgar, Stenness, Maes Howe, Skara Brae).

C. Lewis, Outer Hebrides (Callanish).

D. Colonsay and Oronsay, Inner Hebrides (Cnoc a' Charragh, Scalisaig, Garvard, Kilchattan, The Strand, Oronsay Priory).

E. Jura, Inner Hebrides (the Paps of Jura).

F. Islay, Inner Hebrides (Nave Island, Ardnave Point, Cultoon).

G. Kintyre Peninsula (Campbeltown, Macrihanish, Pubal Burn).

H. Argyllshire (Kilmartin Glen, Temple Wood, Kintraw).

I. Loch Fyne (Brainport Bay, Minard).

J. Glasgow (Knappers, Clydebank; Sighthill; Kilsyth 10 miles northwest of Sighthill).

K. Largs, Ayrshire (The Three Sisters).

L. Arran, Firth of Clyde (Machrie Moor).

M. Prestwick Airport, now Glasgow Prestwick International Airport (HMS Gannet Naval Air Station).

N. Keswick, the Lake District (Castle Rigg).

O. County Meath, Eire (Newgrange, Knowth, Dowth, The Bend of the Boyne; Mound of the Hostages, Tara; Loughcrew).

P. Llangollen, North Wales (Eisteddfods).

Q. Spaceguard Centre (Observatory and modern stone circle), Knighton, Powys.

R. Preseli Mountains, South Wales.

S. Stonehenge and Durrington Walls, Wiltshire.

T. Avebury, Wiltshire.

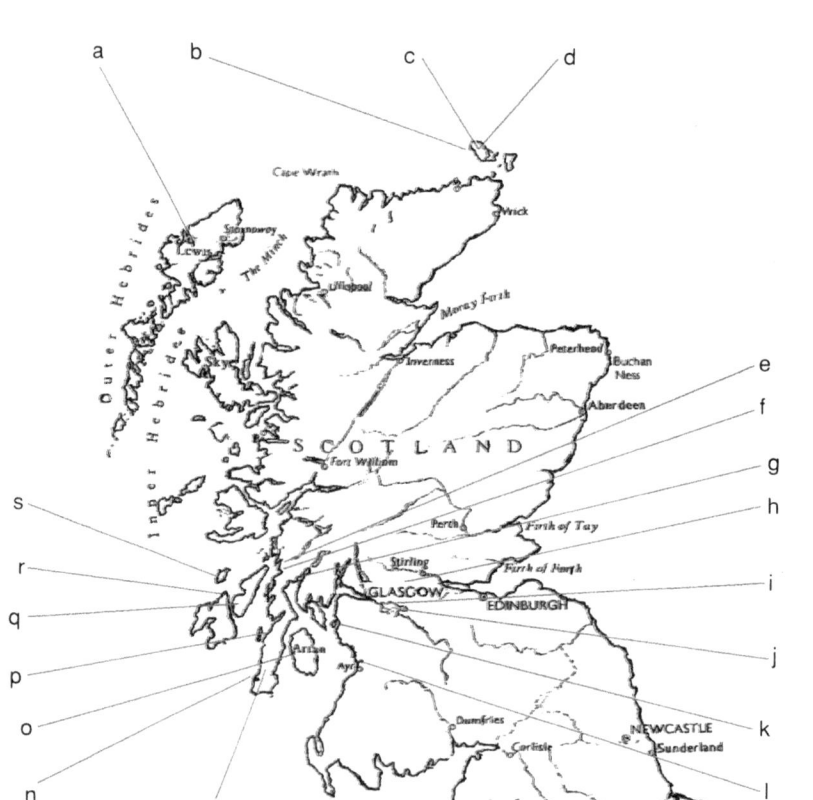

a

b

c

d

Cape Wrath

Wick

Outer Hebrides

Stornoway

Lewis

The Minch

Ullapool

Moray Firth

Inner Hebrides

Skye

Inverness

Peterhead

Buchan Ness

Aberdeen

S C O T L A N D

Fort William

e

f

Perth

Firth of Tay

g

s

Stirling

Firth of Forth

h

r

GLASGOW

EDINBURGH

i

q

p

Arran

Ayr

j

o

Dumfries

NEWCASTLE

k

Sunderland

n

Carlisle

l

m

Key to Map of Scotland

a. Callanish standing stones, Island of Lewis, Outer Hebrides.
b. Skara Brae Neolithic village, Bay of Skaill, Orkney.
c. Ring of Brodgar, Brodgar Peninsula, Orkney.
d. Stenness Stones and Maes Howe chambered cairn, Loch of Stenness, Orkney.
e. Kintraw standing stone and observing platform, Loch Craignish, Argyllshire.
f. Temple Wood standing stones, circle and chambered cairns, Kilmartin Glen, Argyllshire.
g. Brainport Bay alignments, Loch Fyne, Argyllshire.
h. Beltmoss Quarry, Kilsyth.
i. Former Knappers Neolithic complex, Clydebank.
j. Sighthill stone circle, Sighthill Park, Glasgow.
k. Brisbane Glen Observatory and 'Three Sisters' meridian pillars, Largs, Ayrshire.
l. HMS Gannet Naval Air Station, Prestwick Airport, Ayrshire.
m. Former stone circle, Pubal Burn, Kintyre Peninsula.
n. Macrihanish airfield, formerly Royal Air Force, now privatized.
o. Machrie Moor stone rings, island of Arran, Firth of Clyde.
p. Ballochroy standing stones, Kintyre Peninsula.
q. Beinn an Oir, Beinn Tarsuinn, the Paps of Jura, island of Jura, Inner Hebrides.
r. Nave Island and Ardnave Point, island of Islay, Inner Hebrides.
s. Islands of Colonsay and Oronsay, Inner Hebrides – for details see Figs. 4.23A and 4.23C.

Preface

Reasons for writing the book now: the first lunar and stellar observations on site, compiled in the last 4 years, and plans to complete and renovate the circle.

The Glasgow Parks Department Astronomy Project (full title "Astronomy in the Parks") ran from March 1978 to the end of December 1979. In 1978, it had a staff of four people, expanded to ten in 1979 under the Special Temporary Employment Programme, headed by Duncan Lunan (Manager), the late John Braithwaite (Technical Supervisor), Gavin Roberts (Art and Photographic Supervisor) and the late Jean Coles (Secretary). In 1979, the 'Astronomy and Space Education Programme' team included Donald Sutherland (site-tester; on loan to Glasgow University Observatory), the late Bill Braithwaite as model maker, Alan Maclean, Ann Clark and the late Ian Downie as speaker-demonstrators, Richard Robertson as draftsman and Dave McClymont as illustrator.

In collaboration with the Association in Scotland to Research into Astronautics (ASTRA), the Scottish Arts Council and the Third Eye Centre, the Project created the exhibition 'The High Frontier, a Decade of Space Research 1969–1979', which was the largest event of its kind held in the UK to date. On tour, it was estimated that the exhibition was seen by 86,000 people. Work in schools and libraries began in October 1978 during National Children's Book Week and continued throughout the following year.[1]

The Project's initial brief was to build an astronomically aligned monument in one of the city's parks, and this was fulfilled with the creation in Sighthill Park, Glasgow, of the first astronomically aligned stone circle in Britain for between 3,000 and 4,000 years. Appropriately, the main phase of planning and construction took a year and a day, as in the best fairy stories, and the circle was completed in its initial form at the spring equinox of 1979, by a Royal Navy Sea King helicopter.

Following the change of government in the UK in 1979, work was stopped and four stones remain unused; when the circle was landscaped into the park, the plans were misread, and the profile of the circle was spoiled. Meanwhile, solar observations at the circle revealed insights into the functioning of the ancient sites.

In 2006–2008, observations were made for the first time at a lunar standstill event, adding more insights, and the spread of efficient lighting in the city allowed star observations for the first time. A resurgence of interest in the circle in 2001 and again in 2010 has restarted plans to have it renovated and completed, and the time to tell its story has come.

It was welcomed by John Braithwaite, who went on from the project to become the only telescope maker in Scotland, consultant to the public observatories and an inventor of global standing. Sadly, John developed liver cancer in mid-January 2012 and within 3 weeks he was gone. I've paid my tributes to him elsewhere;[2] he and I agreed the dedication of this book long since, and I've let it stand as he would have wished.

Duncan Lunan

References

1. Lunan, D., Ramsay, W.: Astronomy, spaceflight and education. Space Education **1**(11), 488–489 (1986)
2. Lunan, D.: Obituary: John Braithwaite, Telescope Maker. The Herald, February 16, 2012; Lunan, D.: Dedication, in With Time Comes Concord and other stories. Brain in a Jar, Glasgow (2012)

About the Author

Duncan Lunan was born in Edinburgh in 1945 and educated at Marr College, Troon, and at Glasgow University, gaining an M.A. (Honors) in English and philosophy, with physics and astronomy as supporting subjects. He holds a postgraduate Diploma in Education.

He has published four nonfiction books on astronomy and spaceflight to date, plus a collection of short stories and an anthology which he edited. He has contributed to 20 other books to date, and published 33 short stories and nearly 800 articles. He was the assistant curator of Airdrie Public Observatory 1987–1997, and again in 2003–2008. With his wife Linda, he has launched a new organization to support newcomers to astronomy, 'Astronomers of the Future.'

In 1978–1979, as manager of the Glasgow Parks Department Astronomy Project, he designed and built the first megalithic observatory in Britain for 3,000 years; in 1990–1991 he was photo archivist for the Press Centre during Glasgow's year as European City of Culture. In 2006–2008 he was manager of the North Lanarkshire Astronomy Project, organizing nearly 700 events including more than 450 school visits, and otherwise has been a full time writer since 1970.

By the Same Author

Man and the Stars
New Worlds for Old
Man and the Planets
(Edited) Starfield, Science Fiction by Scottish Writers
(Fiction) With Time Comes Concord and Other Stories
Children from the Sky: A Speculative Treatment of a Mediaeval Mystery
Incoming Asteroid! (book coming next year)

Acknowledgements

At the time of the Glasgow Parks Astronomy Project and the first draft of this book, I was married to Linda Joyce Lunan, and my editorial consultant was Paul Barnett, who tried hard to place the book at that time. In April 2010 I married Linda Taylor, who found the advertisement that put me in touch with John Watson of Astrobooks, who pitched it to Maury Solomon of Springer, and I remain deeply grateful to all of them for their help and support. To avoid the confusion of having two 'Linda Lunans' in one book I've reluctantly omitted some photographs and referred in places to 'my first wife,' but not due to lack of appreciation.

For the use of their photographs I'm grateful to Paul Barham, The Boyne Valley Tourist Board, Kate Braithwaite, Tony Crerar, Bill Donald, the late Ian Downie, Brian Fair, Graham Gardner, John Gilmour, Mary Goulder, Bob Graham, Linda Lunan, Euan MacKie, Mark McLaughlin, Frank O'Neill, Julian Paren, Chris O'Kane, Gavin Roberts, the *Springburn Herald*, Mark Runnacles, Chris Stanley and Sigurd Towrie. For permission to use technical drawings from their books, I'm grateful to Alexander Thom, Archie Thom and Euan MacKie, and for others to Jim Barker, Alan Evans, Nick Portwin, to Dave McClymont (Illustrator) and Richard Robertson (Draftsman) of the Astronomy Project for their work, and to Bill Jones, Special Projects draftsman, whose initial plan of the circle (large, and on photo-sensitive paper) would now be difficult to reproduce. For help in processing slides, thanks to Jackson of Glasgow School of Art and to Bob Graham of the North Lanarkshire Astronomy Project.

The Glasgow project's first year team members, the late John Braithwaite, Gavin Roberts and the late Jean Coles saw it through to completion, with growing support in the second year from the late Bill Braithwaite, Ann Clark, Ian Downie, Alan McLean and

Donald Sutherland, plus Dave McClymont and Richard Robertson as above. The project was also supported throughout, and with volunteer help at key times, by the members of ASTRA, the Association in Scotland to Research into Astronautics Ltd, particularly Bill Ramsay and Paul Benson.

In the Glasgow Parks Department (now Land and Environment Services), key players included Tom Bradley (Manpower Services Commission Liaison), Keith Fraser (Director), Ronnie Gray (Principal Landscape Architect), Paul Green (Special Projects Manager), Mr. Hastie (transport), Mr. McKenzie (Architecture and Related Services), Alan Montgomery (Special Projects Operations Manager), Frank O'Neill (Special Projects Shop Steward), Pia Pisaneschi (Special Projects Secretary) and Dave Sneddon (Methane Digester Project). The list is far from complete. Special thanks go to Ian Clair of James Cunning and Cunningham; Mr. Motherwell and his foreman Jimmy at Beltmoss Quarry, to all the other quarry-masters who helped or tried to help, and to Jimmy the Foreman at Sportsworks with his staff, especially the specialist JCB driver; to the late LAEM David Proffitt, RN; and to Lieutenant-Commander Fraser Hutchinson, Lieutenant McBride and Lieutenant Norman Leask, all then of 819 Squadron, HMS Gannet, Prestwick; with further thanks to the Royal Navy Flag Officer at Yeovilton. I must also thank Tony Crerar and Dr. Euan MacKie for keeping me up to date with their work over the years, especially to Euan MacKie for help and support in completing the book.

Further afield, thanks to the late Chris Boyce; Jack Foley; Bob Low of *The Daily Record*; Ben Bova and Kathleen McAuliffe of *Omni*; and Dr. E.C. (Ed) Krupp, Director of the Griffith Observatory, Los Angeles, and editor of *The Griffith Observer*, where parts of Chaps. 8 and 9 first appeared in different form. I've also drawn on my general article 'A Stone Circle for Glasgow' which first appeared in *Tournaments Illuminated*, courtesy of the editor Susan J. Evans; it was later updated for *R.I.L.K.O. Journal*, courtesy of Sylvia Franchetti (now Sylvia Ezen), and excerpts appeared in *Club Meg, The Journal for Stone Circle Builders*, Winter Solstice, 2011, in 'Sighthill Stone Circle: a 33-Year Saga,' a collaboration with the editor, Rob Roy. For the IBM Heathrow Conferences and for the 1982 aerial archaeology flight, special thanks to Leslie Banks, who also provided the aerial photographs of Arbor Low,

Avebury and Stonehenge, initially for a joint presentation by Chris Boyce and myself ('Computers Past and Future'), at the launch of the IBM PC in Renfrew.

"A History of Ancient Britain", by Neil Oliver©, Neil Oliver 2011, appear by permission of The Orion Publishing Group, London.

Finally, as in the dedication, special thanks to the late Professor Alexander Thom, and also to the late Dr. Archie Thom, to Dr. Euan MacKie and to Professor Archie Roy, but for whom none of it would have happened.

For the restoration and completion of the stone circle, help and donations are welcome – contact us at www.sighthillstonecircle.net.

Contents

Part I
Background to the Controversy

1. Archaeologists Versus Astronomers

I looked out over the Moss and tried to imagine life in those days, almost four millennia ago, 200 generations of men away from us. From air photos, I knew that apart from the stone circles, the Moor was covered with hut circles; the whole area must have supported a largeish community of fishermen, hunters and primitive farmers. And at night, the only illumination would be the stars and the Moon and the fires before the huts. No TV, no adverts.... Perhaps for the first time I began to appreciate the appeal these stones, the long cultural echoes of the ancient people of Arran, could have.

– Archie Roy, *Deadlight* [1].

In archaeology, the idea that ancient standing stones and circles were astronomically aligned – and accurately so – is extremely controversial, if not rejected outright. Among astronomers, if not universally accepted, the idea is at least treated with greater sympathy. Members of the public, who are perhaps more generally interested in things astronomical than in the relatively staid work of archaeologists, may be surprised to learn that prehistoric astronomy is not viewed by all scientists as proven fact. Many people may believe, wrongly, that the Druids built Stonehenge; guess that they are up to 3,000 years old; may not even know that there *are* other such sites; and yet are in no doubt at all that Stonehenge and the pyramids were observatories of some kind. In arguments about "gods from outer space," or UFOs, it's virtually taken for granted that the ancient sites relate to the sky.

The opposing views gained prominence in 1967; 10 years later, the argument had not been settled, but was continuing with growing heat. In 1980, fuel was still being added to the flames. Meanwhile the Glasgow Parks Department's "Astronomy in the Parks" project was not intended as a contribution to the debate so

D. Lunan, *The Stones and the Stars: Building Scotland's Newest Megalith*, Astronomers' Universe, DOI 10.1007/978-1-4614-5354-3_1, © Springer Science+Business Media New York 2013

much as a comment on it, by a small group of people who definitely did support the astronomical interpretation of the ancient sites. To date, that comment has not influenced the debate – its point is perhaps a subtle one – but it may have an effect in time. Meanwhile, ongoing events at the circle have provided lessons on what could and could not be done at the ancient sites.

At Stonehenge, in the 1960s, the literature available at the nearby souvenir shop without exception placed some emphasis on the alignment of Stonehenge to midsummer sunrise, while stressing that there was no evidence to support modern names for features such as the "Altar Stone" and "Slaughter Stone [2]." There was nothing to counteract the impression that Stonehenge had functioned at least partly as an observatory.

Behind the scenes, however, controversy was building up. In 1963 C.A. Newham and Professor Gerald Hawkins, working independently, had published interpretations of alignments at Stonehenge indicating a lunar observatory function as well as a solar one [3]. As shown below, even approximate lunar alignments would indicate a much longer and more painstaking program of observations than the solar ones require. Hawkins had used a computer to analyze the positions of the stones, and perhaps that rather unnecessary sophistication made his account the focus of the controversy. Looking back on it in 1984, under the heading, 'Whatever happened to megalithic astronomy?', Christopher Chippindale wrote, "[His] hushed reverence for the wonderful IBM computer which could decode the language of the ancient Stonehenge computer, reads like a period piece now. It is a revealing fact that IBM did not do the first decoding at all: Hawkins did it with good old-fashioned pencil and paper and only checked his result with the machine afterwards [4]."

In 1965 Prof. Hawkins published an enlarged version of his account, with anecdotes and speculation, in his book *Stonehenge Decoded* [5]. Archaeologists at once joined battle – "Moonshine on Stonehenge" was the title of one review. Revising his 1959 HMSO *Stonehenge: Official Guide-Book*, R.S. Newall wrote that year: "The orientation of ancient monuments is not popular with some archaeologists, but, if it be a fact that two stones or one stone and a space between two other adjoining stones are in line with a certain sunrise or sunset, the student may justifiably consider theories based on such facts [6]."

Fig. 1.1 Professor Alexander Thom at his home in Dunlop, Ayrshire (Photo by Gavin Roberts, 1978)

Alexander Thom, professor of Engineering Sciences at Oxford (Fig. 1.1), had by that time devoted 40 years of study to megalithic sites in Britain, much of it with the assistance of his son, Dr. Archie Thom. His ideas as to the purposes of the ancient builders were based on personally conducted surveys at nearly 400 megalithic sites. So far, however, his published results had appeared only in mathematical journals, except for his first paper on the subject, which had appeared in the *Journal of the British Astronomical Association* in 1954. Otherwise, his work was known only to acquaintances such as Archie Roy, who had published his own survey of the Arran site in 1963. So far, Thom's effect on the archaeological world had been minimal; but now the lid came off. Entering the controversy in 1964 with articles in *New Scientist*, Thom followed up with summaries of his work in book form under the titles *Megalithic Sites in Britain* (1967) and *Megalithic Lunar Observatories* (1971) [7]. They could hardly have been more different in tone from *Stonehenge Decoded*, but their opponents treated them as all of a kind.

From Thom's studies he concluded that the ancient builders had been engaged in a long-term program of lunar observations, tracing the Moon's complex movement in the sky with a thoroughness and precision going far beyond the needs of the calendar or of any imaginable ritual. Along the way it seemed that the megalithic astronomers would have realized that Earth is a sphere and discovered the slow wobble of Earth's axis, which we call the precession of the equinoxes. Since many archaeologists were already committed to the position that the Stonehenge alignments were coincidental, Thom's thesis met with widespread hostility. It was perhaps unfortunate that at that time the Thoms' surveys had not so far included Stonehenge itself: critics tended to lump all the astronomical theories together, as if arguments against one applied to all, whereas in fact there were very important differences between Thom's claims and those made by Hawkins, followed by Sir Fred Hoyle [8].

Within the relatively compact boundaries of Stonehenge, Hawkins had allowed the postulated astronomical alignments to be wide of the mark by as much as a degree and a half, either way, altogether six times the apparent diameter of the Sun or the Moon; however, the sites that Thom had studied were alleged to show very careful exploration of the landscape, lining up astronomical events with features on the horizon to an accuracy of a few minutes of arc. And whereas Hawkins and Hoyle maintained that Stonehenge was a computational device, used to predict eclipses, Thom was claiming that the sites he surveyed were the product of an observing program, locating each alignment precisely on the ground before it was immortalized in stone. The sophistication which then followed, in the layout of the various types of stone rings, was pure mathematics – based on right-angled triangles and inspired by a wish to make the key dimensions multiples of a standard unit of length.

The hostile reaction of the archaeologists to the new discipline – generally called "astroarchaeology" or "archaeoastronomy," according to the background of the speaker – was directed to both sides of the question. The astronomy involved was alleged to be far too subtle for Neolithic people involved in a daily struggle for survival, possessing no metal tools, instruments or even the art of writing. But the principles involved are not that hard to grasp (see below) and the mathematics are only of first-year undergraduate level.

One can sympathize with an archaeologist with a liberal-arts background who finds them hard going, but not if the objection takes the form that "ancient Man couldn't have done anything I don't understand."

Far more to the point, however, is that the calculations are needed only to set back the clock of Earth's shifting axis, back to prehistoric times. The Neolithic observer did not need to calculate where the Sun rose at the solstice of 2700 B.C., because he could watch it! It is no coincidence that megalithic astronomy is widely accepted by amateur astronomers, who spend their nights watching for meteors or aurorae with the naked eye or with small telescopes. To the prehistoric Britons, occupying a sparsely populated island thousands of years before the invention of the street lamp, with generally better weather due to heightened solar activity between 3000 and 2000 B.C. [9], the sky overhead was anything but an abstraction.

Referring to the star lore of pre-Columbian Central America, and its possible survival in Amerindian cultures, Michael D. Coe wrote: "This area has hardly been touched, and the reasons are not hard to find. In the first place, there is scarcely an ethnologist or social anthropologist who can identify anything other than the Moon or the Big Dipper in the night sky; the so-called natives are a great deal wiser [10]."

Just two examples may serve to underline that point. In Aubrey Burl's comprehensive book *Prehistoric Avebury* (1979), he cites and accepts some evidence of astronomical alignments, but goes on: "That the rituals involving human bone at Avebury were held in the darkness of the night is unprovable but the fact that its two Coves, themselves in the likeness of tomb entrances, were connected with the night, the North Circle Cove facing towards the moon's most northerly rising, the Cove at Beckhampton towards the sunrise at midwinter, suggests that nocturnal winter activities were performed here.... Whether, then, from the west along the Beckhampton Avenue or southwards from the Sanctuary, one can imagine torchlit processions as the Moon rose..." which he goes on to imagine in detail. But the Moon rises at its most northerly position, not at every midwinter, but only once every 18.61 years. For about a year either side, the Moon would rise near enough to that position for ritual purposes – but often in daylight, seldom if ever at the Full, and not at midwinter unless by pure chance, and even then, not to be repeated for centuries.

"The restricted size of these Coves means that they could never have been used for scientific observation of the heavens [11]." No; but to orient them with sufficient accuracy to be recognizable, despite the changes in the sky since then (Chap. 2), the movements of the Moon would have to have been observed and recorded over several 18.61-year cycles – even with better weather, probably over more than a century – before the builders could risk pinpointing its northerly rise with a stone structure.

More blatantly, a recent TV documentary began with the bold statement, "Stonehenge has NOTHING to do with astronomy" – then treated the viewer to reconstructions, based on recent archaeology, showing feasts, processions and rituals at Stonehenge and nearby Durrington Walls, held at the summer and winter solstices, along pathways and avenues aligned with the solstice sunrises and sunsets [12]. It would give the impression that finding those dates, and determining those alignments, was either done by magic or was so easy that it couldn't be classified as astronomy. As the later chapters of this book will show, it's never as easy as that, even if you know approximately what the answers are supposed to be.

The other attack on Thom's work was based on the archaeological side of the question, but was carried out with what – to an outsider – seemed remarkable confidence, given that the megalithic builders did not leave written evidence. Crudely put, the argument was that Neolithic Britain did not have the social structure needed to support an astronomical program. Until recently, there was no evidence for anything more advanced than small tribal communities under warrior kings who ruled by sheer physical supremacy. But, since there were no written records, nor even representative drawings, to assert that the kings and their priests *can't* have been interested in astronomy, and that the astronomical alignments of their monuments must be coincidental, seemed to be going altogether too far. There was already counter-evidence at the time of the Glasgow Parks Astronomy Project [13], and more recent discoveries in the Stonehenge area, in Orkney and in Argyllshire, among others, have acted to radically change this position (see Chap. 4).

Fig. 1.2 Dr. Euan Mackie in his study at Glasgow University, Hunterian Museum (Photo by Gavin Roberts, 1978)

One archaeologist who did have misgivings about the attacks on Professor Thom was Dr. Euan MacKie, Assistant Curator at the Hunterian Museum of the University of Glasgow (Fig. 1.2). MacKie had previously been involved in research concerning the Mayan culture of Central America. The Maya, and their predecessors and successors, the Olmecs and Aztecs, did maintain a priesthood with a keen interest in the calendar and the movement of the stars and planets; their temples, cult centers and in some cases entire cities appear to be astronomically aligned. But almost all the literature of the Maya and Aztecs was destroyed by the conquistadores in great bonfires ordered by their priests. There was so little left that the significance of the astronomical passages was a matter of some doubt. Were it not for the surviving calendars, and sculptures showing astronomer-priests at work, there would be no proof that the alignments of Mayan structures are anything but coincidental.

In Peru and Bolivia, where the pre-Incan and Incan cultures were like the Mayan and Aztecs in many ways, there was no writing, and so even less is known. But in Britain we don't even know whether the megalith builders were conquered or not, let alone have testimony from the conquerors about the astronomical activities of their predecessors! We just don't know who built the megaliths, or why they stopped; until recently there was nothing to go on – except the remains of pottery, weapons, some large wooden structures, some rubbish tips and the megaliths themselves.

After immersing himself in the astronomy and satisfying himself that Thom did have a strong case, MacKie turned to the megaliths to look for supporting evidence. Of the various digs he conducted (see Chaps. 4 and 9), the one at Kintraw in Argyllshire is perhaps the most significant. On the hillside overlooking the great standing stone MacKie found, as predicted by Thom's analysis, an artificial stone platform from which the stone lined up with a prominent "V" between the peaks of Jura, 28 miles away, marking very accurately the sunset at midwinter solstice around 1800 B.C. (Figs. 1.3, 5.4).

In *Science and Society in Prehistoric Britain* and *The Megalith Builders*, MacKie went on to argue that Neolithic society had supported a class of "professional" astronomer-priests like those of the Maya [14]. He suggested that the great wooden henges such as Durrington Walls had been the homes of the astronomers, not ritual centers as was generally supposed; the anomalous quantities of particular bones in the associated rubbish tips would show, not that they were used only for ceremonial feasts, but that the dwellers were supported by surrounding communities, the slaughtering and butchering being done elsewhere.

The houses at Skara Brae in the Orkney Islands, built and furnished in stone because wood was scarce, would be surviving examples of the conditions once enjoyed at Woodhenge and Durrington Walls. The flat-bottomed "Grooved Ware" associated with the wooden henges would be the special pottery of the inhabitants, used on their tables and shelves, whereas the round-bottomed ware of the common people was intended to sit on earth floors and right itself if jostled.

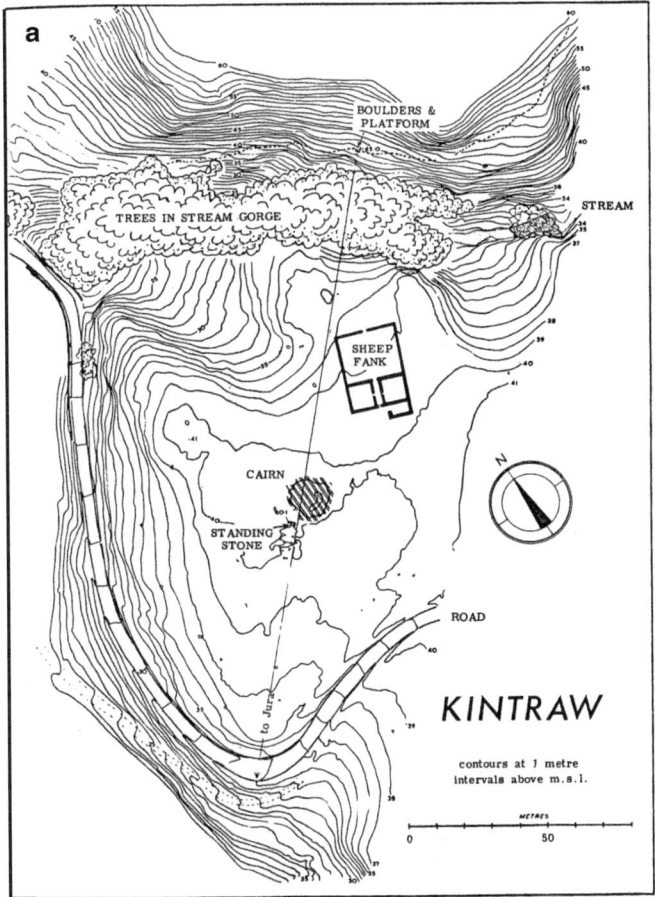

FIG. 1.3 The alignment to Jura from the platform at Kintraw. After Euan MacKie, "Science and Society in Prehistoric Britain [14]"

To an outsider, the hostile reaction to MacKie's thesis was hard to understand. For example, in *The Prehistoric Settlement of Britain* Richard Bradley wrote: "There is also convincing evidence for mathematical and astronomical expertise (Thom 1967). This evidence has been sorely abused (e.g., MacKie 1977), but there can be little doubt that a fifth class must now be envisaged: a priesthood, or its equivalent, whose role cannot be rationalized in terms of subsistence agriculture... [15]." Just what the objection to MacKie's work might be was not explained. The reference is to *Science and Society in Prehistoric Britain*, mind you, and not to the more speculative and controversial final chapter of *The Megalith Builders*, in which MacKie suggested that the idea of a professional priesthood may have spread along the Mediterranean from a very early origin in Mesopotamia, and come to Britain, with the art of building chambered tombs in stone, from Portugal. Mackie bases the suggestion on an apparently common unit of length measurement used in the appropriate areas, and it is only a suggestion.

Simultaneously caricaturing the argument and engaging in an astonishing piece of "lumping together," Prof. Glyn Daniel, the editor of *Antiquity*, wrote in the July 1980 issue of *Scientific American*: "The number of books on Stonehenge and on other megalithic monuments...is also, alas, an all too clear demonstration of the imagination, wishful thinking and credulousness of many authors, and the abysmal ignorance of many alleged archaeologists who can only be styled, if uncharitably, as fantasy buffs.... As recently as 1977 MacKie in *The Megalith Builders* declared that they were the work of wise men from predynastic Egypt and Sumeria. There are others, among them Erich von Däniken, who see the megalith builders as voyagers from space. Now there is also a widespread belief that these monuments were built with an astronomical purpose, and such words as 'astroarchaeology' and 'archaeoastronomy' are freely bandied about... [16]."

Who could guess from that paragraph that what MacKie actually suggested is that the *idea* of a professional priesthood might have originated in Mesopotamia, and that the suggestion comes very tentatively at the end of two entire books on the British megalith builders? There is no discussion here or later in Daniel's article of any of the points in MacKie's argument for the existence of an astronomer-priest caste in Neolithic times; MacKie's name is mentioned only to provide a guilt-by-association link between the astroarchaeologists ("the Hawkins-Thom bandwagon," as Daniel classifies them below) and the von Däniken books, whose inaccuracy and sensationalism have been rightly attacked elsewhere. The Thoms' brief reply concluded with the words, "Finally, sir, in publishing this *ad hominem* attack, you have shown the spirit of your publication to be neither scientific nor American."

That letter was not selected for publication, but in reply to another protest, the editors allowed Glyn Daniel to add, "the greater number of people who study prehistoric megalithic monuments in Europe remain unconvinced by the work of Thom, Hawkins and others who often write with a profound ignorance of the archaeological facts [17]." MacKie held his fire, but subsequently took mild exception to being placed by Daniel in the ranks of 'lunacy' and 'deluded men' [18, 19].

Such was the background of contention against which 'Astronomy in the Parks' took place in 1978–1979, and when the first draft of this book was prepared in 1982. It's gratifying to be publishing it just when the prevailing view of Neolithic society has changed out of all recognition.

References

1. Roy, A.: Deadlight. John Long, London (1968)
2. Atkinson, R.J.C.: Stonehenge and Avebury, and Neighboring Monuments. Ministry of Public Buildings and Works. Her Majesty's Stationery Office, London (1959)
3. Newham, C.A., quoted in Emmott, D.: The Mystery of Hole G, Yorkshire Post, 16 Mar 1963; Hawkins, G.S.: Stonehenge decoded. Nature 200, 306–308 (1963)
4. Chippindale, C.: Life around Stonehenge. New Sci 1404, 12–17 (1984)
5. Hawkins, G.S., White, J.B.: Stonehenge Decoded. Souvenir Press, London (1965)
6. Newall, R.S.: Stonehenge, Wiltshire, Fourth Impression, Amended. Ministry of Public Buildings and Works, Ancient Monuments and Historic Buildings. Her Majesty's Stationery Office, London (1959/1965)
7. Thom, A.: Megalithic Sites in Britain. Clarendon Press, Oxford (1967); Thom, A.: Megalithic Lunar Observatories. Clarendon Press, Oxford (1971)
8. Hoyle, F.: On Stonehenge. W.H. Freeman, San Francisco (1977)
9. Eddy, J.A.: The case of the missing sunspots. Sci Am 236(5), 80–88 & 92 (May 1977)
10. Coe, M.D.: Native astronomy in mesoamerica. In: Aveni, A.F. (ed.) Archaeoastronomy in Pre-Columbian America, pp. 3–31. University of Texas Press, Austin and London (1975)
11. Burl, A.: Prehistoric Avebury. Yale University Press, London (1979)
12. Secrets of Stonehenge. Yesterday TV, UK, 25 Aug 2011

13. Anonymous: Science and the citizen: getting the axe. Sci Am 238(1), 69 (1978)
14. MacKie, E.: The Megalith Builders. Phaidon Press, London (1977); MacKie, E.W.: Science and Society in Prehistoric Britain. Elek, London (1977)
15. Bradley, R.: The Prehistoric Settlement of Britain. Routledge and Kegan Paul, London (1978)
16. Daniel, G.: Megalithic monuments. Sci Am 243(1), 64–76 (1980)
17. Daniel, G.: Letter. Sci Am 243(5), 8 (1980)
18. Daniel, G.: Editorial. Antiquity 55, 81–89 (1981). Quoted in MacKie, E.W.: Implications for archaeology. In: Heggie, D.C. (ed.) Archaeoastronomy in the Old World, pp. 117–139. Cambridge University Press, Cambridge (1982)
19. MacKie, E.W.: The prehistoric solar calendar: an out-of-fashion idea revisited with new evidence. Time Mind: J Archaeol Conscious Culture 2(1), 9–46 (2009)

2. Now You See It, Now You Don't

How great is the power in the intersection of circles!.
– Astophon, *Mineralium Constellatorum*, quoted
by Giordano Bruno [1].

Imagine that the Sun is a perfectly clear transparent globe, and that an observer is stationed at its heart. It would be a remarkably peaceful place to be (neglecting what would happen if the Sun *really* became transparent). All around there would be stars, so far away that they would seem to be fixed to the inside of a sphere that rotated very slowly around the observer with a period of just over 25 days, as the Sun itself rotated on its axis. It would be obvious that the band of the Milky Way ran right around the sky and was much thicker on one side than on the other; if the observer had a sufficiently accelerated time-sense, so that millions of years passed in what seemed to us like seconds, then he or she would realize that all the stars were circling around an invisible center of mass in the thickest part of the Milky Way. Before the first 230-million-year revolution was over, he might have deduced that the visible part of the system was disc-shaped, and that the Sun's orbit lay well out towards the edge of the disc.

If, however, his time-sense was a thousand times slower than that, the changes would seem much more gradual. The Sun's motion relative to the surrounding stars is a mere 12 miles/s, and the local scenery would change much more slowly. He would realize, as the thousands of years ticked past, that the stars near Vega were gradually separating one from another, and that those on the other side of the sky were gradually drawing together. A few stars would show perceptible motion. Arcturus, for example, would be seen creeping towards the southwest corner of Boötes. The configuration of the Plough, a.k.a. the Big Dipper, would gradually

D. Lunan, *The Stones and the Stars: Building Scotland's Newest Megalith*, Astronomers' Universe, DOI 10.1007/978-1-4614-5354-3_2, © Springer Science+Business Media New York 2013

change in a way that showed that most of its stars were traveling together. The observer might even realize that some of the other bright stars spread around the sky shared the same motion, so that the Sun was passing through a widely scattered cluster. (That was big news, early in the twentieth century [2].) But most of the star background would seem to be motionless, over thousands upon thousands of years.

Slow down by another factor of a thousand, however, so that single years go by in subjective seconds, and the planets would be visible – although Mercury and Venus would be moving so fast that it would be hard to see them as moving points rather than flickering bands. But it would be obvious, even for the more distant planets, that each one moves around the Sun in a fixed plane, relative to the starry background.

Earth, trailing across the field of vision, would not be very conspicuous – in the model Gavin Roberts planned, with the Sighthill circle representing the Sun and the orbit of Pluto set at Glasgow's city boundary (Chap. 10), Earth was less than an inch across and 200 yards out! The Moon, smaller and less reflective, would be still harder to see, but its flitting around Earth would be obvious – above, ahead, below, behind, and not repeating on a year-to-year basis. On this timescale, and assuming no abnormal powers of vision, the Moon would be the only visible object with a continuously changing path.

Earth and Moon form a twin-planet system. The Moon is so far out that the perturbing pull of the Sun would wrench it away from us, but the combined gravitational attractions of the Earth and Moon are enough to keep them together. Strictly speaking, however, the Moon does not orbit around Earth. It orbits the Sun in a path that is continuously modified by Earth into a series of scallop-shaped curves [3]. Seen from the Sun, the Moon would move continually forward against the starry background. If it were close enough to Earth, and therefore moving fast enough to move "back," seen from the Sun, it would be moving too fast for the human eye to follow, on this timescale.

Supposing that Earth, too, were a transparent airless sphere, and that the viewpoint moved to its center; then the timescale would have to be slowed by another factor of a thousand to make events comprehensible. Even then it would seem that the day lasted about 3 s, with the Sun, Moon and stars wheeling continuously around, parallel to the equator. It would be obvious at once that the Moon was moving from west to east, against the starry background, and swinging north of the equator, then south, over the minute or so which it took to go around the sky and complete its cycle of phases. Even on the second time around, it should be obvious that the Moon's path against the stars had slipped against the stars. It might even be obvious that the Sun's pull was literally dragging the Moon back, forcing the orbital plane to rotate.

The other factor at work, the pull of Earth's equatorial bulge, would be harder to deduce. Meanwhile the Sun's movement would become apparent, spiraling up the heavens to the Tropic of Cancer at northern-hemisphere midsummer, then south again to reach the Tropic of Capricorn at the midwinter solstice. It might take quite some time to realize that the Sun, too, was moving west to east against the stars, and the apparent movement up and down the sky was a result of the tilt of Earth's axis. Since the major clues would be lacking at the center of a transparent Earth, the observer might never realize that Earth was going around the Sun, not vice versa.

Over a year – subjectively, about a quarter of an hour – the slippage of the Moon's orbit would be very apparent. It takes only 18.61 years to precess right around the sky. The corresponding effect, of the pull of Sun and Moon on Earth's equatorial bulge, would take longer to show, but eventually it would become clear that, while the Sun's path against the stars remained unchanged, the whole reference frame of equator, tropics and poles was tracing a very slow circle against the stars. That effect, the precession of the equinoxes, takes 26,000 years for a single cycle; by the time he recognized it our imaginary speeded-up observer might realize that the axial tilt itself was diminishing, bringing the tropics very slowly closer together. Over 5,000 years the midsummer Sun would have moved south by about its apparent diameter, and the midwinter Sun northward by the same amount.

Suppose, finally, that our observer moves up to Earth's surface, that Earth becomes opaque, and that the atmosphere arrives to complicate matters – displacing objects from their true position in the sky by the effect of refraction. And suppose that our observer's time sense is slowed by more than 30,000 times, to the normal rate of human perception. Now it takes 4 min for Earth's rotation to change the position of something in the sky by a single degree. How could a true understanding of such slow processes be reached? Indeed, what could move one to the attempt?

Well, first there is the alternation of day and night. Awareness of that cycle is programmed into us at a very basic level – in isolation, without external clues, human beings tend to drift towards a 20 h rhythm that goes back 600 million years, to the time when life crawled from the sea onto the land. It is hard to say whether the dominant factor here was the day/night cycle or the ebb and flow of the tides. But it is the braking effect of the tides that has slowed down Earth since then, and it is the perception of the day/night cycle that resets our biological clocks.

The monthly orbit of the Moon around Earth has its biological tie-ins, too. The link with the menstrual cycle has been noted from very ancient times. Even today, for all the barriers we have put between nature and ourselves, outbreaks of crime and admissions to asylums tend to peak around the full Moon – for no obvious reason. The Moon does affect the weather, as the old proverbs say, although the effect is so subtle that it took computer analysis of a century's records to prove the point.

However, to the nomadic hunters who comprised the human species for almost all of its existence, the phases of the Moon were crucial. The amount of light to be expected after sundown was a key survival factor, to be taken into account at the start of the day, and there are notched bones to suggest that the monthly cycle was being plotted 20,000 years ago or more. It must have seemed fortunate indeed that, when the days were shortest and game hardest to find, the Moon rode higher in the sky and shone brighter for longer.

The evidence suggests that the beginnings of agriculture did not come directly from the annual cycle of plants but from the movements of the great herds of game. From moving with the herds, as the Lapps do even today, domestication and nomadic herding comprised the next step and led to the first attempts at agriculture and fixed settlements. The very oldest towns, such as Jericho (c. 8000 B.C.), came before the first crop farming. The move to an annual cycle obviously required a true calendar, even if none had previously been attempted.

The year is harder to calibrate than the month, however. Even today, a week is a long time in politics – and most societies, if not all, tried to fit their lunar calendar into the year. The cause is a lost one. The lunar and solar cycles are not commensurate; i.e., they do not fit together in any straightforward numerical relationship – and different cultures made different compromises to divide the year into approximate "months" of convenient length, while keeping religious ritual and agricultural practice in step with the solar year.

Aubrey Burl says of the Neolithic era, when stone-age farming still coexisted with coastal hunter-gathering and nomadic herding, "It was a time when there were no months or weeks [4]." A clearer example of modern humanity's divorce from the sky would be hard to find; and yet that divorce is extremely recent, for many people still within living memory. An 1869 star atlas in the author's collection shows the full panoply of stars, month by month, from due south of St. Paul's in London, and another book shows the brighter stars from Westminster Bridge, month by month, above the 1930 gas lamps. In 1926 H.V. Morton wrote of "the Plough flinging its clear symbol over a powdered sky" above the Embankment [5]; and in one old Encyclopedia you can find a plate of the stars from Blackpool beach, without the Illuminations, with the Tower silhouetted against the natural background glow of the sky. Even in this era, being able to navigate by the stars has gotten this author out of trouble three times, including one memorable occasion in downtown Los Angeles.

To review the events discussed above from a ground-based viewpoint, Fig. 2.1 shows the celestial sphere as viewed by an observer in the northern hemisphere. The altitude of the pole above the northern horizon is equal to the observer's latitude, and the heavenly bodies circle around it, parallel to the equator, with the daily rotation of Earth.

The *altitude* of a body above the horizon, and its *azimuth* measured along the horizon from the north point, change constantly as Earth turns. Apart from the circumpolar stars, which are too near the pole to rise and set, everything else rises in the east and sets in the west at a position that is determined by the *declination* of the object, measured from the equator (Fig. 2.2). Where the declination equals the observer's latitude, the star passes overhead once a day.

Without metals or metal tools to make the mural quadrants and other sophisticated instruments of pre-telescopic astronomy,

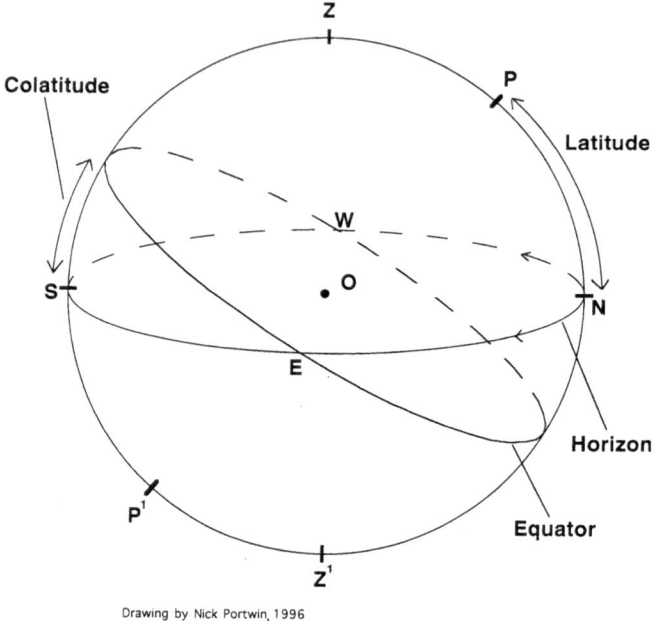

Drawing by Nick Portwin, 1996

FIG. 2.1 Altitude-azimuth coordinates (Drawing by Nick Portwin)

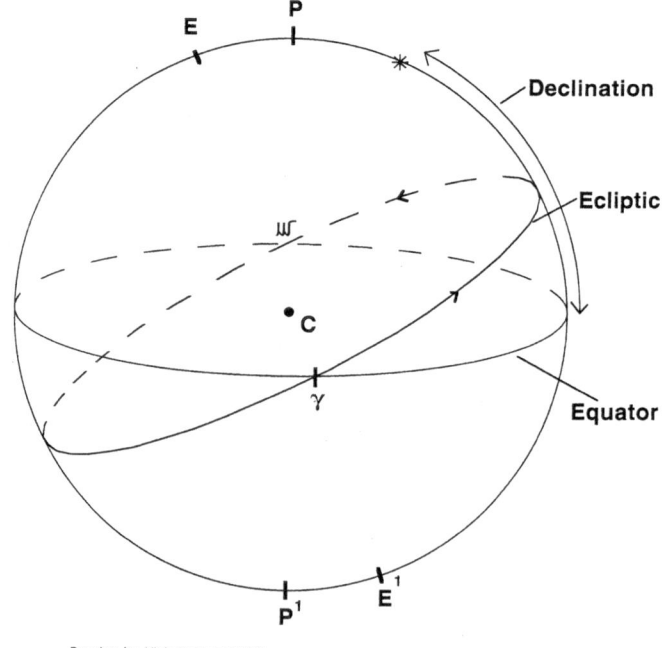

Drawing by Nick Portwin, 1996

FIG. 2.2 Right ascension and declination coordinates (Drawing by Nick Portwin)

positional horizon observations were all the Neolithic builders could do. The plane of Earth's orbit around the Sun is the *ecliptic,* the center line of the zodiac; as the Sun moves along it over the course of the year, its horizon position varies from its most southerly midwinter rise and set, when it's overhead at the Tropic of Capricorn, to its most northerly midsummer rise and set, when it's overhead at the Tropic of Cancer. It's universally agreed that the Stonehenge Avenue and the later structure both mark the midsummer sunrise.

The plane of the Moon's orbit is inclined to the ecliptic by approximately 5°. Under the pulls of the Sun and Earth's equatorial bulge, the Moon's orbital plane swings around in the sky, against the background of stars, with a period of 18.61 years (regression of the lunar nodes). When the Moon reaches its highest possible position above the ecliptic, it rises and sets at its extreme northerly positions, and at its extreme southerly positions 14 days later. These events are termed the Major Standstill. As we've seen, few archaeologists agree with Gerald Hawkins that the 'station stones' of Stonehenge I mark the extreme northern and southern positions of the Moon's 18.61 year cycle; and still fewer with Alexander Thom, that the megalith builders had a sophisticated program of lunar observatories, spread over the British Isles.

Just about 9.3 years after the Major Standstill, the Moon comes to an intermediate position called the Minor Standstill, flanking the solstice risings on the other side. This was easiest to illustrate using the planetarium at Glasgow Science Centre. A friend of the author's, Chris O'Kane, stood at the midsummer sunrise position on the edge of the dome, while we advanced the projector setting by 6 months, and I then stood at midwinter sunrise position. Then the operator brought the Moon to the Major Standstill northerly rise, on Chris's right; and we advanced 14 days to the southerly rise, on my left. Then, advancing 9.3 years, we repeated the exercise for the Minor Standstill, with the northerly moonrise now to Chris's left and my right. There was no time to practice beforehand, and it was a great relief to see the Moon rise and set at those positions.

When it comes to star alignments, the position is more complex. Because Earth's equatorial plane doesn't coincide with the ecliptic, nor with the orbit of the Moon, the combined pulls of the Sun and Moon on Earth's equatorial bulge cause Earth's axis to 'wobble' around the ecliptic pole with a period of 26,000 years

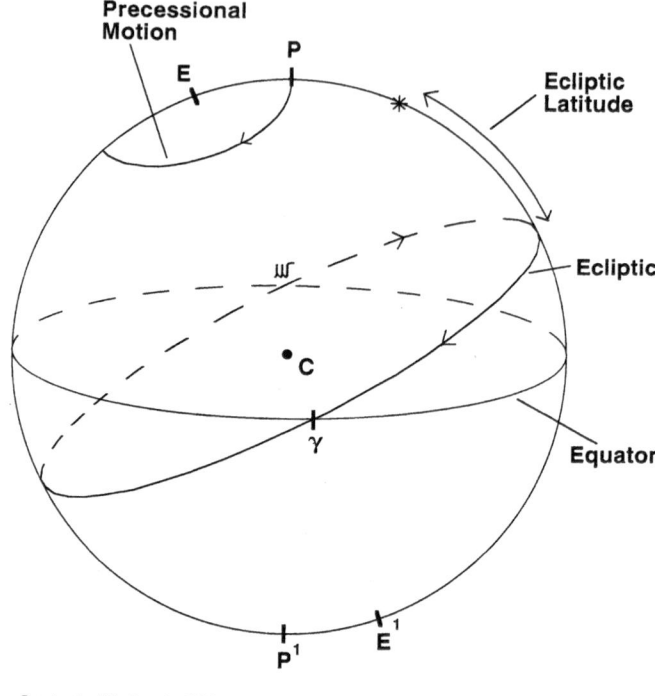

Drawing by Nick Portwin, 1996

FIG. 2.3 Ecliptic coordinates (Drawing by Nick Portwin)

(Fig. 2.3). About 13,000 years ago the north pole star was Vega in Lyra, and 5,000 years ago it was Thuban in Draco, at the time of Stonehenge I. The pull makes the equator move around the ecliptic, constantly changing the position of the vernal and autumnal equinoxes (precession of the equinoxes, first described outside the Arab world by Robert Grosseteste in the thirteenth century). As a result a star's declination is constantly changing – likewise its right ascension, which is measured from the vernal equinox along the equator, in the same direction as the Sun's motion on the ecliptic shown by the arrows in Fig. 2.2.

When the pyramids were built the Pole Star was not Polaris but Thuban, the brightest star in Draco. Polaris, which passes closest to the pole very soon now, has been regarded as 'fixed' in the sky for most of the last 1,000 years – Shakespeare has Julius Caesar compare himself to it for constancy – though Columbus realized it had to be making a small circle in the sky relative to the new-fangled magnetic compass "..for the needle moveth not"!

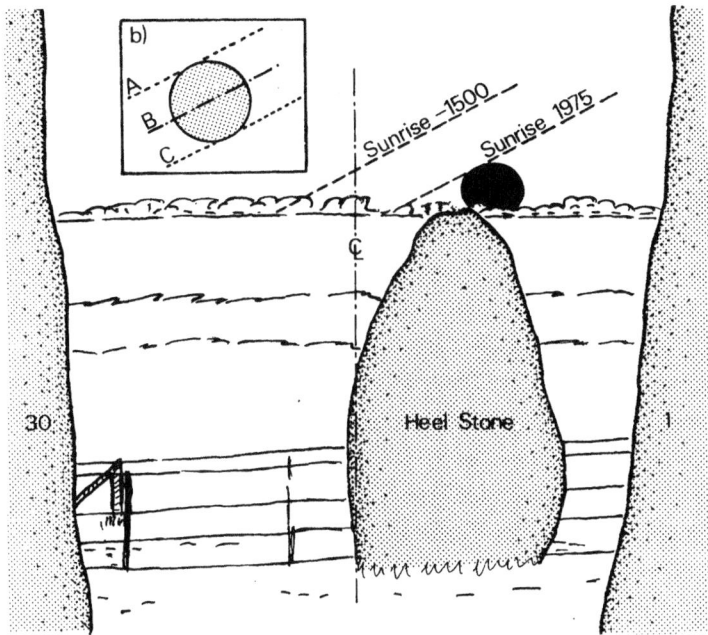

FIG. 2.4 The changing position of Stonehenge midsummer sunrise (After Peter Lancaster Brown) [6]

But even precessional change varies slightly with time, because the tilt of Earth's axis relative to its orbital plane (the obliquity of the ecliptic) changes gradually. When Stonehenge I was built the Obliquity was very nearly 24°, but it has now declined to 23° 27'. One effect of this is that the Sun now rises over the Heelstone on Stonehenge Avenue, but in Neolithic times it rose to the left, on the center line of the avenue (Fig. 2.4).

The Moon's standstill rising and setting positions have altered similarly. But in addition to that, other forces, such as the pull of the planets, bring about multiple, small, superimposed changes in the tilt of Earth's inclination and of the Moon's orbit, all altering its rising and setting positions on the horizon; and Thom maintained that all these had been detected by the Neolithic observers, and recorded in multiple alignments at complex lunar observatories such as Temple Wood at Kilmartin in Argyllshire.

Astronomers can partly get around the problem of coordinate change by giving star positions in ecliptic latitude, which remains nearly constant, and ecliptic longitude, which changes smoothly

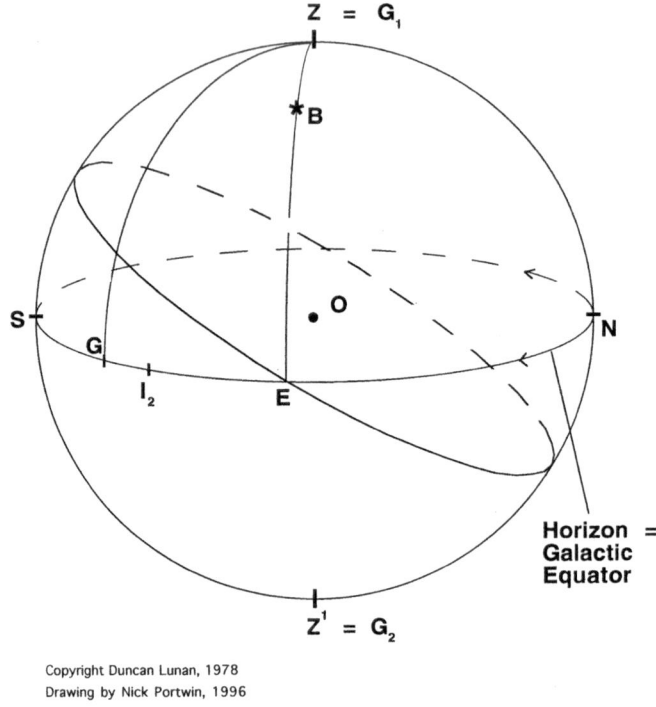

FIG. 2.5 Galactic coordinates (Drawing by Nick Portwin)

with time. But for astronomical coordinates that are fixed over time spans longer than human lives, we have to use galactic latitude and longitude, whose zero point is the galactic center and whose pole lies on the perpendicular to the plane of the Milky Way (Fig. 2.5). The Sun takes 200–230 million years to orbit the galactic center, so for all but the nearest stars, galactic co-ordinates are effectively fixed even beyond the spans of civilizations. You might suppose that galactic coordinates could have no part in megalithic astronomy, but even that seems not to be entirely true – see Chap. 5.

References

1. Yates FA. (1964) Giordano Bruno and the Hermetic Tradition. Routledge and Kegan Paul, London
2. Turner, H.H.: A Voyage in Space. Society for Promoting Christian Knowledge, London (1915). Includes two maps showing the Sun's position within the cluster from different angles

3. Firsoff VA. (1959) Strange World of the Moon: An Enquiry into Lunar Physics. Hutchinson, London

4. Burl, A: Prehistoric Avebury, op cit.

5. Dunkin, E.: The Midnight Sky: Familiar Notes on the Stars and Planets. The Religious Tract Society, London (1869); Phillips, Rev. T.E.R. and Steavenson, Dr. H.W. (eds.): Hutchinson's Splendour of the Heavens, Hutchinson, London (1930); Morton, H.V.: London by Night (1926). Reprinted in: H.V. Morton's London, Methuen (1940)

6. Brown, P.L.: Megaliths, Myths and Men: An Introduction to Astro-archaeology. Blandford Press, Poole, Dorset (1976). A similar illustration appears in Krupp, E.C.: The Stonehenge Chronicles. In: Krupp, E.C. (ed.): In Search of Ancient Astronomies. Chatto and Windus, London (1979)

3. Ancient Astronomy Around the World

We're here, we did this, this is not nature
But geometry, see it from the moon some day!
> – Prof. Edwin Morgan, *Planet Wave: The Great Pyramid (2,500 b.c.)*

In the last 50 years there have been amazing changes in our knowledge of the Solar System, the universe and the past – even our own. In the 1960s it was thought that there were no varieties of human older than a million years, and that *homo sapiens* went back only 40,000; the history of all previous human forms was summed up in Ashley Montagu's title, *Man, His First Million Years* [1]. Even at the end of the decade, in a huge book commemorating the Moon landing, editor David Thomas wrote, "*Moon* treats of a vast number of events through what seems, to this editor at least, a vast period of time – nearly 4,000 years. This period, of course, represents but an infinitesimal fraction of the million-year-long evolution of the human race" [2].

By the late 1970s it was recognized that *homo sapiens* went back at least 90,000 years [3]. By the late 1990s that became possibly 130,000 or more [4], maybe even 300,000 [5]. By 2007 it was 200,000 years for sure, with the move from the savannah to the coast back to 164,000 years [6]. In Nigel Kneale's *Quatermass and the Pit* (1958–1959), suggestions that hominid life went back three to five million years were called 'wild guesses,' but now we know that earlier forms of humankind date back at least five million years, maybe even seven million.

Throughout that time, human beings have lived in a much closer relationship with the sky than we can easily picture today. The earlier, nocturnal phase of our existence has left us with excellent night vision, especially on the periphery, but throughout the

D. Lunan, *The Stones and the Stars: Building Scotland's Newest Megalith*, Astronomers' Universe, DOI 10.1007/978-1-4614-5354-3_3, © Springer Science+Business Media New York 2013

hunter-gatherer, nomadic and early agricultural phases of society, awareness of the movements of the Sun and the Moon has been essential. Pace Aubrey Burl (see Chap. 1), there has never been "a time when there were no months or weeks," and probably we have been embroidering on that awareness, in storytelling and in art, for a very long time indeed.

Some of the oldest records of this may survive in Australia. Very recently, the possible remains of an astronomically aligned stone circle have been found at Mount Rothwell, 50 miles from Melbourne, dating from around 8000 B.C. – contemporary with the first human settlement in Britain after the last Ice Age. But humans have been on the continent for at least 30,000 years – some say 60,000–76,000 [7] – and although claims for 176,000 years failed to stand up, there was settlement of Flores in Indonesia by hominids 700,000 years ago or more, proving the ability to make short sea crossings [8]. There's growing recognition that some legends of 'the Dreamtime,' before the coming of white explorers, have a basis in fact – for example, stories in northeastern Australia of 'a great white wave' may relate to a mega-tsunami caused by a meteorite impact off New Zealand in the fifteenth century [9]. For a long while, it's been noted that the Aboriginal name for the Henbury meteorite craters, formed around 2700 B.C., translates as "sun trail fire devil stone" [10].

In recent years increasing attention has been paid to the star lore of the Aborigines, including 'dark constellations' – dark nebulae of dust and gas, silhouetted against the brightest regions of the Milky Way, which Aboriginal legends portrayed as the Emu, the totemic emblem of many tribes [11]. Binyu, the Ringtail Possum, featured as the head of the Southern Cross, while the Coalsack Nebula behind it was a dungeon into which he hurled his wife's seducer [12]. Eta Carinae, nearby, dominated the constellation of Cullowgillouric War, the wife of War, the Crow; Sirius was the brightest star in Warepil, the Wedge-tail Eagle; Achernar the brightest in Yerredetkurrk the Fairy Owl; and the Magellanic Clouds were Kourtchin, 'the Brolgas' [13]. It's suggested that these constellations may correspond to figures in rock paintings dating back to 40,000 B.C., and if so, they may be the oldest known portrayals of the sky. The South African bushmen and tribes in the Amazon basin have similar 'dark constellations' [14]. They and the Quechua of Peru saw the Emu as several dark constellations, including two llamas being attacked by a fox.

Still more recently there's evidence of an advanced Amazon culture, with towns of up to 60,000 people – wiped out so rapidly by diseases from Europe that its towns, which were oriented to the cardinal points on the horizon, were so quickly overgrown by the jungle that few explorers ever got to see them [15]. There may be big revelations in archaeoastronomy to come there: one of the Amazonian dust constellations is the Rainbow Serpent, of whom we will say more below.

The oldest known artistic materials, kits including an ochre-rich mixture in abalone shells, were found in a cave on the South African Cape Coast and dated to around 98,000 B.C. [16]. The oldest known human art, on clay pieces more than 70,000 years old, was also found in South Africa, at the Blombos Caves [17]. One of the oldest known astronomical depictions is thought to be a carving of the Moon's phases, on an engraved bone plaque from the Dordogne, around.27,000 B.C.; and the next oldest a similar tally carved on a bone tusk at Gontzi, in the Ukraine, 15–10,000 B.C.

The most famous ancient paintings of all, at Lascaux in France (c. 17,000 B.C.), are in a cave that is dome-shaped and illuminated by the Sun only at summer solstice. Of 130 known caves of the period decorated with paintings or carvings, only four are not aligned with solstices or equinoxes, while the orientation of undecorated caves appears random. The Lascaux sequence matches the zodiacal constellations with the Bull, a unicorn representing Capricornus and horses fitting Sagittarius [18]. The figure representing Taurus appears to show the position of Aldebaran, between us and the Hyades cluster, as it was in 15,300 B.C. due to the precession of the equinoxes, while rows of dots below the horse and stag figures may represent the lunar cycle [19]. The evidence for truly ancient astronomy may be all around us, in both hemispheres, much of it still unrecognized.

As far as we know, civilization emerged in the Middle East around 10,000 years ago, around such sites as Jericho and Ebla. Around 5000 B.C. a new culture emerged at the far end of the Mediterranean, in southern Spain and northern Africa. Spreading along the Mediterranean, its members created stone circles

FIG. 3.1 The sarsen uprights and lintels of Stonehenge III, the last great stone circle project in the British Isles. Vistamorph™ photo by Chris O'Kane

and megalithic temples at least as far as Malta, and there are arguments that it reached Asia Minor. Moving north, the same people spread into Brittany, around the coasts of the British Isles and on into Scandinavia. It seems they entered the British Isles around 4500 B.C., following earlier hunter-gatherer and Mesolithic settlement around 11,000 B.C., at the end of the Ice Age. In Ireland they built huge mounds, surrounded by standing stones, at the Hill of Tara and in the Bend of the Boyne, most famously at Newgrange. It's long been known that above the entrance passage into the Newgrange mound, a window in the stonework admits the light of the rising midwinter Sun. Recent excavations have revealed that there are similar passages at the nearby mounds of Knowth and Dowth, aligned to sunrises and sunsets at the solstices and equinoxes; and in Great Britain, the Hebrides, Orkney and Shetland they built stone rings, rows, and simpler standing stones, culminating around 1800 B.C. with the final phase of Stonehenge at the beginning of the Bronze Age (Fig. 3.1) (see Chap. 4).

Meanwhile, back in Sumeria, around 4000 B.C., a new art form had emerged. With written documents still in the form of clay tablets, people were putting their identifying marks on them with 'cylinder seals': devices like little paint rollers that could be run over the wet clay to leave a strip of recurring symbols and images. When followers of Immanuel Velikovsky insist that there is no record of the planet Venus prior to 1500 B.C., it's worth pointing out that there is an unbroken, evolving sequence of Venus images, representing her as Inanna and later as Ishtar, from 4000 B.C. to the start of the Christian era [20]. Also, from the outset, they show figures in the landscape, human and divine, and over them shapes in the sky that we recognize – the Lion, the Bear, the Bull, the Plough and the long Dragon.

There's nothing inevitable about the patterns. They look very different in different latitudes, and in ancient China for instance the sky was divided up in quite a different way. Still nearer the equator the tendency is not to form figures at all but 'ropes' of stars running in parallel lines across the sky from east to west. Such 'star-ropes' were later used by the Polynesian cultures to achieve their amazing feats of navigation between islands widely separated in the Pacific [21]. It's interesting that in ancient Sumerian cosmology, the model of the universe was an equatorial one – a half-cylinder running north to south, unkindly compared to a Nissen Hut by the late Arthur Koestler [22]. The implication is that some of them, at least, must have lived much nearer the equator at a still earlier date, and indeed standing stones with lunar and stellar alignments have been found far to the south in the Rift Valley (though apparently much more recent, dated to 300 B.C.) [23]. Yet the Sumerians' constellations had the forms we know. In Mesopotamia, they were far enough north to see the stars wheeling around the celestial pole, and indeed the constellation Draco – in which the pole was located at the time – features on cylinder seals with the Plough, the Lion and the Bear.

What we now call the classical constellations came to us from ancient Greece, with a considerable input from the Arabs, who were the foremost astronomers meantime. Although the Greeks devised the versions which we still use today, much of their star lore came largely from Egypt – but not all of it. In a thesis entitled *The Lamps of Atlantis* Prof. Archie Roy of Glasgow University argued that the major clue was a poem, "The Phenomena," written by Aratus in 250 B.C., putting into verse a description of the celestial sphere by Eudoxus a century earlier. The remarkable thing is that the description of the constellations fits the Mediterranean of 2,000 years before; in the meantime precession had shifted the pole and the equator a long way against the stars. Eudoxus and Aratus describe stars that couldn't be seen from Greece or Egypt in their time, and leave out stars that were then visible; from that, the compilation date and the latitude of the sphere's creation can be deduced. Eudoxus obtained the sphere in Egypt – "a discontinued line" as Roy put it – and probably they obtained it from the seafaring Minoan civilization [24].

The sphere dated from around 2800 B.C. and was made on the latitude of Crete and Thera. If the Egyptians had done it, they would have put in stars further south, ones that they could see from Egypt. Archie Roy suggested that the site might have been Santorini (formerly Thera), the volcanic island that sustained an advanced culture before it was destroyed in an explosion about 1450 B.C. Thera may well have been the inspiration for the Atlantis legend – hence Roy's title – as well as the Biblical story of the Exodus [25]. It was a worse disaster than Krakatoa, in its initial violence and its consequences, since the tidal waves were produced within the waters of the Mediterranean, and survivors from the island civilizations may have founded the cities in Greece and Asia Minor that were to fall out a few centuries later in the Trojan War. However, ancient Egypt, primarily an inland civilization, escaped largely unscathed, and Roy argued that they fell heir to the Thera model of the sky. His colleague, the late Prof. Michael Ovenden, considered that the most 'important' constellations, the 12 houses of the zodiac within which the Sun, Moon and planets move (plus Ophiucus, the Serpent-Bearer), were designed when those figures were 'upright' with respect to the equator – about 2700 B.C.

There's another clue. The polar constellations of the Egyptians were different, featuring a hippo, a crocodile and the leg of an ox, not the Lion, Bears, Dragon and Boötes, the Herdsman, as we know them, nor the Plough. Henri Frankfort, who compiled a respected reference work on cylinder seals in the 1930s, thought these figures couldn't be constellations because Draco, the Dragon, is so faint. He didn't realize that in those days Draco housed the pole and the sky turned around it! When he took that into account, Roy noted that Draco also encloses the ecliptic pole, around which the Sun's apparent motion is centered. Draco must have seemed much more important when it housed both the major hubs of the sky. And since the major civilization of the time in the right latitude was the Sumerian culture of Mesopotamia, whose early art showed the figures of the polar constellations that we know, Roy changed his thesis and ascribed the drafting of the constellations to them [26].

Around 3000 B.C. Mesopotamia also saw a major new form of architecture: the evolution of the Ziggurats, stepped pyramids oriented to the cardinal points, north, south, east and west. This was after the creation of the great stone circles in Orkney, soon after the building of Newgrange in Ireland and not long before work began at Stonehenge I. Other cultures, too, had their eyes on the sky. Apart from the Plough (the Big Dipper, in the United States) and Orion's Belt, most people can recognize the Pleiades or "Seven Sisters," above Orion to the right. In 3000 B.C. they were on the celestial equator, rising due east, and are so described in the Sanskrit *Satapatha Brahmana*. But, fascinatingly, the Aryan culture was not then in India but still concentrated in the Middle East – at the place and time that the constellations were being devised [27]. The Pleiades were then very close to the vernal equinox, and maybe they continued to be associated with them until the Aryan entry into northern India, between 1500 and 1200 B.C. We still call the vernal equinox 'the First Point of Aries,' as a nod to ancient Greek astronomy, and most astrologers still treat Aries as the first sign of the zodiac, although the equinox has been in Pisces for most of the Christian era – hence the fish as an early Christian symbol – and the musical play *Hair* notwithstanding, it has not yet reached Aquarius.

Nearly 3,000 years after the Ziggurats, the emperor Ch'in, who united China (named after him), built the Great Wall and was buried with his army of terra-cotta figures, ordered a great burning of books because he claimed to have rediscovered the ancient knowledge and found later texts to be corrupt. Surviving texts date the origin of the world to 2850 B.C. (contemporary with Stonehenge I) and the origin of astronomy to 2500 (contemporary with completion of the Great Pyramid of Giza). Ch'in himself was buried in a great pyramid made of earth, which has still to be opened.

An American missionary named Geil, who walked the Great Wall in the early 1900s collecting legends of its origin, wrote that the stories told nearest the center of the empire – perhaps the most reliable – held that the Wall was built as an image of something in the sky [28]. Geil considered the idea that the Wall and the dragons of Chinese art both represent Draco, but like Henri Frankfort later, he thought it couldn't be true because Draco is too faint. He failed to realize that between 2850 and 2500 B.C. the pole star was

Thuban, the brightest star in Draco. Geil himself suggested that the wall represents the Milky Way, which certainly is prominent in Chinese star-lore, but not in a way that makes ancient knowledge differ from the new.

The Sumerians and their successors continued to build using mud bricks, but elsewhere an architectural revolution was at hand. The oldest civilization in Peru was a coastal culture that raised great platforms in stone, aligned to the cardinal points, around 2700 B.C. This was contemporary with the building of Silbury Hill and the great stone circle of Avebury, with its great serpentine Avenues, in England – when the pole star was Thuban. At Tucume a complex of 26 of these raised platforms is the largest concentration of pyramids in the world, where the late Thor Heyerdahl made his home and conducted digs in the early 1990s [29]. And in Egypt, around 2650 B.C., the priest Imhotep 'taught the Egyptians to build in stone.'

Imhotep undertook the building of the first pyramid, the Step Pyramid of Saqqara, for the pharaoh Djoser (Fig. 3.2a). During construction it went through several changes of form, evolving from the shape of a typical mastaba, a mud-brick tomb structure, into an internally reinforced pyramidal shape. Imhotep claimed that his inspiration came directly from Thoth, the god of astronomy and mathematics. In December 2007 two huge underground chambers were detected at Saqqara, possibly housing the tomb of Imhotep, so there may be sensational discoveries to come [30]. The Pyramid's steps may have been filled with earth and paved with marble to give the appearance of a true pyramid (Fig. 3.2b) [31]. If so, then there may well have been star alignments built into it, as well as being aligned to the cardinal points, like the pyramids of Mesopotamia.

Imhotep didn't pass on the secret of building stable pyramids, and the outer layers of the next one at Maidum collapsed due to an earthquake, causing the slope of the next again to be drastically modified to create the Bent Pyramid of Dahshur. The epoch of building great pyramids in Egypt lasted only 150 years, culminating in the three gigantic pyramids of Giza and the enigmatic figure of the Sphinx (Fig. 3.3). Star alignments built into the Great Pyramid of Cheops (Fig. 3.4) include symbolic sight-lines to Thuban and to Kochab in Ursa Minor, on the northern face, and to Sirius and Alnitak in Orion's Belt on the southern. Robert Bauval argues that

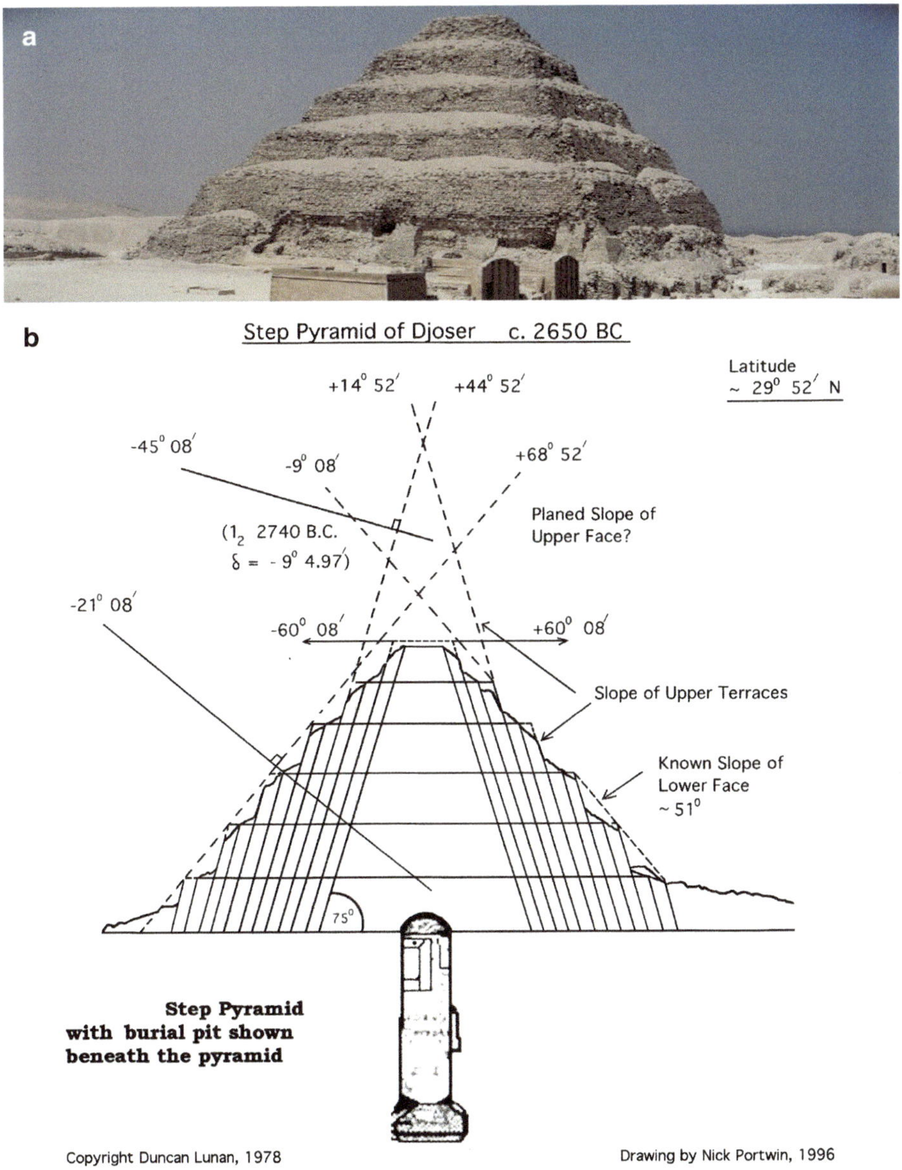

b

Step Pyramid of Djoser   c. 2650 BC

Latitude
~ 29° 52′ N

+14° 52′   +44° 52′

-45° 08′

-9° 08′   +68° 52′

Planed Slope of
Upper Face?

$(1_2$ 2740 B.C.
δ = - 9° 4.97)

-21° 08′

-60° 08′   +60° 08′

Slope of Upper Terraces

Known Slope of
Lower Face
~ 51°

75°

**Step Pyramid
with burial pit shown
beneath the pyramid**

Copyright Duncan Lunan, 1978   Drawing by Nick Portwin, 1996

FIG. 3.2 (a) The Step Pyramid of Djoser at Saqqara. Vistamorph™ photo by Chris O'Kane. (b) Plan of the Step Pyramid (Drawing by Nicholas Portwin)

the entire layout of the pyramids on the Giza plateau, in relation to the Nile, represents the stars of Orion's Belt in relation to the Milky Way (Fig. 3.5) [32]. (More controversially he and Graham Hancock have gone on to suggest that the layout truly represents

FIG. 3.3 The Sphinx with the pyramids of Mycerinus at *left*, Chephren *behind* and Khufu (Cheops) at *right*

FIG. 3.4 The Great Pyramid of Khufu (Cheops)

FIG. 3.5 The three pyramids of the Giza plateau. Vistamorph™ photos by Chris O'Kane

the configuration of an earlier 'First Time' around 10,500 B.C., and that the Sphinx dates back to it [33]. That takes us into arguments about the age of the Sphinx [34], and leads them to postulate a world-spanning civilization otherwise unknown to us – issues outside the scope of this book.)

Fɪɢ. 3.6 (a, b) Solar alignment in the Temple of Amun-Re, Karnak, near Luxor. Photos by Chris O'Kane

Astronomy in ancient Egypt was far from over, however. The evolution of the solar temple at Karnak (Fig. 3.6) and the star maps first known in the tomb of Hapsheptut, most famously the one known as 'the Dendera Zodiac,' are too well known to need description here. But Chris O'Kane, inventor of the Vistamorph™ lens through which he took the remarkable panoramic photographs of this chapter and the next, believes that the planet Mars has a central role in Egyptian cosmology that has still to be recognized. To this day Egypt's capital, Cairo, takes its name from Al Kahir, the Conqueror, the Arabic name for Mars.

While mentioning controversies, it's often asked whether it can be mere coincidence that the Ziggurats, the pyramids of Tucume and Egypt, Stonehenge, Avebury, and the origins of Chinese (claimed) and Mayan civilization (see below), all lie within the same 500 years. In their book *The Cosmic Serpent*, Victor Clube and Bill Napier suggested that a 'super-comet,' similar to the icy bodies Phoebe and Chiron in the outer Solar System, might have broken up among the inner planets around 3000 B.C., filling the sky with comets and meteor showers to trigger a new, global interest in the sky [35].

At least two impact events, and possibly three, then and after, might support the idea. Alan Bond and Mark Hempsell reckon that a low density 1 km rock asteroid with a mass of 800 million metric tons passed over Sumeria at 14 km/s on June 29, 3123 B.C., clipping the Gamskogel ridge and impacting the Köfels area in the Austrian Tyrol, triggering a massive landslide that erased the main crater, while smaller fragments caused other impact features in the area. "As the object travelled up the Adriatic Sea... and across the Alps the supersonic shock would have caused considerable destruction on the ground beneath the trajectory. The impact... would release energy equivalent to 1.4×10^{10} t TNT. [The plume] would rise... to some 900 km before falling over the Levant and Sinai causing considerable destruction over a wide area.... There would have been many direct casualties, near 100 % mortality over areas of thousands of square kilometers in both the Alps and the Near East. There would also have been a severe global climate change that caused further death and social disruption" [35]. The formation of the Henbury craters near Alice Springs in Australia (see above) is in the right time period, but the meteorite fragments from there are nickel-iron, implying that the object came from the Asteroid Belt, rather than the Kuiper Belt or the Oort Cloud, which are the sources of comets. The third event, which takes us to astronomy in the Bible, is Noah's Flood.

The oldest surviving account of the Flood is in the Sumerian "Epic of Gilgamesh" and dates from around 2250 B.C. [36]. Other flood legends are found elsewhere in the world, but differ in major details. Versions of the Gilgamesh story are found in Egypt, the Hittite kingdom, India and China, and the Biblical one – which was picked up by Jewish exiles in the Babylonian captivity – is the only one that leaves out the impact. Not knowing that, Isaac Asimov nevertheless suggested that there had been one, in the Persian Gulf, in his essay "The Rocks of Damocles" [37]. It began with "a cloud no bigger than a man's hand" (a distant mushroom?), then a tsunami ("the fountains of the deep were unleashed," and the Ark was carried inland to Ararat), and only after that the sky grew dark and "the windows of heaven were opened" with torrential rain (Genesis 7, 11). But the Sumerian account has a heat flash, an incandescent rising cloud with ejecta ("the Annunaki lifted up their torches"), a ground shock, an air blast, and only then the tsunami and the deluge.

Further off, a Hittite legend says the flood was caused by the Moon falling to Earth (descending fireball), and the Egyptian account says it began with fire from the constellation Leo, while divine personages stalked the land striking down the populace with iron maces. The ancient Egyptians knew iron only from meteorites, and the Leonid meteors still provide spectacular displays every 33 years. It's been suggested that parts of the story of Samson are a confused account of a Leonid fire-storm, although other writers associate him with Orion and its January meteors (see below) [38]. Between 2354 and 2345 B.C. there was an abrupt turndown in global climate, and there is now evidence that the impact may instead have been in the Iraqi marshes, only a century before the Gilgamesh text [39]. It's remarkable that one nineteenth-century estimate put the date of Biblical deluge at 2349 B.C., though it's probably a coincidence because other Bible studies then put it much further back [40].

There may be another impact event in the Bible. In Psalm 18, after David calls to God for help: "Then the earth shook and trembled; the foundations also of the hills moved and were shaken, because he was wroth. There went up a smoke out of his nostrils, and fire out of his mouth devoured: coals were kindled by it. He bowed the heavens also, and came down: and darkness was under his feet. And he rode upon a cherub, and did fly: yea, he did fly upon the wings of the wind. He made darkness his secret place; his pavilion round about him were dark waters and thick clouds of the skies. At the brightness that was before him his thick clouds passed, hail stones and coals of fire. The Lord also thundered in the heavens, and the Highest gave his voice: hail stones and coals of fire. Yea, he sent out his arrows, and scattered them; and he shot out lightnings, and discomfited them. Then the channels of waters were seen, and the foundations of the world were discovered at thy rebuke, O Lord, at the blast of the breath from thy nostrils."

Professor Baillie suggests that similar imagery in Psalm 74 and the book of Isaiah relates to impacts, possibly the 2,350 BC event [39]. But the same imagery is found in 2 Samuel 22 and is thought to be contemporary with I Chronicles 21:16: "When David looked up and saw the angel of the Lord standing between earth and heaven, and in his hand a drawn sword stretched out over Jerusalem, he and the elders, clothed in sackcloth, fell prostrate to the ground" – very probably a comet recorded in Chinese annals for the 970s or 960s B.C. [41].

The banners of the nation of Israel's four divisions were constellation figures – Judah, the Lion; Reuben, a man and a river (Aquarius); Ephraim, the Bull; Dan, the Eagle and Serpent [42]. Orion is one of the few constellations mentioned in the Bible, in the Book of Job (9:9 and 38: 31–32), and both times Arcturus and the Pleiades are also mentioned. Charles Herberger suggests that the relationship with the Pleiades at least is no coincidence, because many of the mysteries of the Book of Samson could be explained if he was originally a Cretan Sun-king, represented by Orion, whose legend was brought to the Middle East by the Philistines. The lion killed by Samson would be the constellation Leo, and the bees that afterwards nest in the lion's body would stand in Cretan mythology for the soul of the previous Sun-king, whom Samson had deposed [38]. In classical times Regulus in Leo was one of the four Royal Stars marking the onset of the seasons: Aldebaran for spring, Regulus summer, Antares for autumn and Fomalhaut for winter. Now, due to precession of the equinoxes, they all appear earlier in the year [42].

The 300 foxes with flaming tails, with which Samson destroys the crops of the Philistines, might be the Orionid meteors from Halley's Comet. Earth encounters the shower 300 days after the winter solstice. For the bizarre incident of the jawbone of the ass, used to slay the Philistines and afterwards to provide water, Herberger suggests that the captivity in Egypt may be responsible rather than the later one in Babylon. Egyptian mythology describes the god Set as having an ass's head on Earth, but a bull's head and leg in the sky, identified with the stars we call the Plough or the Big Dipper, part of Ursa Major. (It's often suggested that Job's references to Arcturus are mistranslations of the Bear – but see below.) In the actual constellation of the Bull, however, the open clusters of the Hyades and Pleiades have a long-established association with rain. Delilah, Herberger suggests, was a Moon-goddess; and the pillars which Samson breaks are the solar pillars found in Philistine and Jewish temples.

There's a further possibility, however. Carl Sagan suggested that an exploding star in the cluster Praesepe, 'the Beehive,' might have been part of the Samson legend [43]. In classical maps the tuft of the Lion's tail is sometimes represented as the constellation Coma Berenices, between Leo and Boötes, the Herdsman, the summer counterpart of Orion – and the brightest star in Boötes is Arcturus.

As we leave the Old Testament, in the first millennium B.C. a new age of astronomical observation was opening up in the Far East. In China, Japan and Korea detailed records were being kept of changes in the sky: bright meteors, meteor showers, comets and exploding stars.

There were other cultures, though, who still felt the urge to match the shapes seen in the sky with great structures on the ground. For instance, at least some of the figures "drawn" on the plain of Nazca in Peru, and also found on the local pottery of the time, can be matched up with the stars. The lines and figures have been formed by clearing away the dark pebbles that cover the Nazca plain to reveal lighter sand beneath; they radiate in star-like fashion from a number of centers.

In the 1970s much was made of the fact that a few of the lines point to the former azimuths of rising stars, but those appear in random order and many others don't, with any accuracy better than chance [44]. (This is a good counter to the archaeologists who insist that the real astronomical alignments at other sites 'must' be random.) It now seems that the lines and the so-called 'runways' may link hills and wells, hoping to attract rain during a 40-year drought in the sixth century [45]. But grouped around the ends of the 'runways' there are also huge figures such as a condor, a dolphin (far from the coast) and even more strangely of a type of spider found only in the Amazon jungle, also far from Nazca. These figures are also found on local pottery and the figure of the condor, for example, makes a good match to the constellation we call Pavo, the Peacock [46].

The sky legends of Peru are closely linked with those of Mexico, where the Maya and later the Aztecs worshipped Quetzalcoatl, whose planet was Venus but whose symbol was the winged, feathered serpent. The high point of the Mayan civilization was 200 B.C. (contemporary with the building of the Great Wall) to A.D. 200, and it was thought that they had inherited their astronomy from the Olmec culture before then, but it's been found that the oldest Maya sites in Belize go back to 2600 B.C. [47], contemporary with the pyramid builders, and back to the time when the celestial pole lay in Draco – making the feathered serpent, and the jeweled one in the Amazonian constellations, more interesting still.

The Maya shared Central America with another great civilization, centered at Teotihuacán in Mexico, abandoned around A.D. 750. The grid of the city and its two great flat-topped pyramids is not aligned to the cardinal points, and isn't obviously astronomical, though the sight-lines may have incorporated Sirius and the Pleiades. Other features termed 'pecked crosses' have clear solar orientations.

Despite the wholesale destruction of Mayan records, enough has come down to us in writing or on stone to show that they had an advanced society with a priesthood who practiced astronomy, among their other arts. Drawings and carvings exist of Mayan observers using crossed sticks to conduct observations. A similar exercise to determine the distance of the Moon used to be part of the practical work in first-year astronomy at Glasgow University. The Mayan observations enabled them to construct complex, interactive calendars stretching far into the past and the future. There's been a lot of fuss and even a blockbuster movie about one that comes to the end of a cycle late this year, and by the time this book comes out that issue will be settled – one way or the other, it's wickedly tempting to add. But other predictions extend much further.

The Caracol building at Chichén Itzá in the Yucután even looks like a modern observatory, although that's because part of the two-cylinder structure has fallen in [48]. An initial survey by Gerald Hawkins found that the openings in the structure marked the equinoctial sunsets, and the extreme northern and southern moonsets [44]. Further work by Aveni *et al.* found alignments to midsummer sunrise and sunset, the rising of Castor, Pollux and Canopus, the setting of Fomalhaut, and the northernmost setting of Venus [48].

(That northerly setting is for A.D. 1000 [49]. Venus's *extreme* setting position is the most northerly of any planet in the Solar System, made possible because it's the closest planet to us, and it will be reached in 28,000 years [50]. An alignment to that is incorporated into the 'Sunstones II' sculpture by Richard O'Hanlon and David Cudaback, outside the Lawrence Hall of Science at the University of California at Berkeley. It's a little optimistic to suppose that it will still be accurate then: at the foot of the hill is a football stadium whose two halves have not lined up since the earthquake of 1906.)

The Castillo, the main pyramid at Chichén Itzá, was built with input from the later Toltecs of the coastal region. An hour before sunset at the equinoxes, the shadows and sunlight on the northern balustrade of the stepped pyramid take on the figure of a diamondback serpent, undulating down the pyramid as the Sun descends. By 1983 an estimated 10,000–12,000 people were turning up to see that at the vernal equinox, and these places are now big tourist attractions, as are Stonehenge and Newgrange.

That brings us to a major question on which to close the chapter. What is astronomy? Obviously the unknown designer of the Castillo was motivated to provide a spectacle, probably part of an intense religious experience and now a theatrical event. With that as with many of the ancient sites and practices described in this chapter, a critic may say, 'But that's not astronomy!' Yet to create them required detailed observations, analysis and prediction of the movements of the heavenly bodies – astronomy, and specifically *positional* astronomy, by any normal definition.

MacKie now feels that the emphasis which he and the Thoms placed on the accuracy of the observations, has polarized the argument; has made it seem that they thought the ancient observers pursued astronomy for its own sake, with the same motivations that drive modern science, and without regard to other human concerns [51]. He now feels that his own book title, *Science and Society in Prehistoric Britain* [52], was misleading and attracted some of the misplaced criticism his work incurred. Clearly there's a measure of truth in that. Chapters 4 and 5 in this book will demonstrate that some of the other ancient sites were built for theatrical effect, no doubt for religious and political reasons – just as President Kennedy's call to put a man on the Moon was most likely to promote the interest of the Democrat party and particularly his Vice-President, Lyndon Johnson, in the southern states, while regaining prestige for his administration after the Bay of Pigs debacle, and generating a non-military competition with the USSR to replace dangerous confrontations such as the Cuban missile crisis. But none of that detracts from the scientific achievements of the Apollo program, many of which were accomplished out of the limelight.

MacKie now feels that even his comparison of Neolithic society with the Mayan may have been overstated [51]; but now that we know Neolithic society was extensive and did have a professional priesthood, that retraction may be premature. Chapters 4 and 8 will show that precise observations *could* have been conducted at Neolithic sites, sometimes tracking events that would be of interest only to serious students of the movements of the Moon, and human curiosity being what it is, we may suspect that if they could have done it, they probably did it.

This quick overview of astronomy in the ancient world is very far from complete. North American readers will justly regret the omission of Medicine Wheels and serpent Mounds [!] throughout pre-Columbian America; in Africa, there's no mention of Zimbabwe and its neighbors, to say nothing of peripatetic cities nearer the equator; and in India, pre-telescope positional astronomy reached great heights of achievement. In his contribution to *Cosmic Perspectives*, below, a breathtaking leap by Jean-Claude Pecker dismisses the entire 2000-year Western dead ends of geocentric theory and the fixed, unchanging heavens, not even in a sentence but in a semi-colon [53]. But omissions are inevitable. For fuller introductions to the subject see G.S. Hawkins, *Beyond Stonehenge*, and E.C. Krupp's *Echoes of the Ancient Skies*, below, because the authors have toured the world and visited the ancient sites in person.

References

1. Montagu, A.: Man, His First Million Years. Mentor, New American Library, New York (1958)
2. Thomas, D. (ed.): Moon, Man's Greatest Adventure. Abrams, New York (1973)
3. Mellars, P.: Major issues in the development of modern humans. Curr Anthropol 30(3), 349–385 (1979)
4. Leakey, R., Lewin, R.: Origins Reconsidered, In Search of What Makes Us Human. Little, Brown and Co., New York (1992)

5. Irwin, A.: Skull find pushes back man's origin. Daily Telegraph, March 29 (1997)

6. Highfield, R.: A modern child – but born 160,000 years ago. Daily Telegraph, March 13 (2007); anon, Human odyssey fuelled by mussels. The West Australian, Oct 18 (2007)

7. Dayton, L., Rintoul, S.: Genes map aborigines' arrival. The Australian, Sep 23 (2011)

8. Leech, G.: The first boat people. The Australian Magazine, July 18–19 (1998)

9. Bryant, T.: Tsunami: The Underrated Hazard. Cambridge University Press, Cambridge (2001); Menzies, G., 1434, The Year a Magnificent Chinese Fleet Sailed to Italy and Ignited the Renaissance. HarperCollins, New York (2008)

10. Anonymous: Australia's Henbury Craters. Sky & Telescope 49(5), 287–290 (May 1975)

11. Ford, V.: Mt. Stromlo and Siding Spring observatories: May sky the most spectacular for clusters, nebulae and clouds. The Weekend Australian, May 2–3 (1987)

12. Berry, A.: Aborigines who followed a star. The Sunday Telegraph, 1 Oct, 1995, reporting on Hayes, R.: Dreaming the stars, the astronomy of the Australian aborigines. Interdiscipl. Sci. Rev. 20(3), 187 (Dec 1995)

13. Healy, G.: Good heavens: an Australian zodiac. The Weekend Australian, Oct 24–25 (1998)

14. Ford, V.: Earth's tilt takes us to midwinter's day. The Weekend Australian (June 1987)

15. Unnatural Histories: Amazon. BBC-4 Channel, UK, 24 Jan 2012

16. Anonymous: Art, 100,000 years old. Daily Telegraph, 14 Oct 2011

17. Oldfield, M., Mitchison, J.: Quite interesting: oldest thing. Daily Telegraph, 3 April, 2010

18. Lost Worlds: Prehistoric Astronomers. Channel SBS-1, Western Australia, June 2009

19. Final Frontier. BBC-2 Channel, UK, 12 Jan 2001

20. Frankfort, H.: Cylinder Seals. Macmillan, London (1939)

21. Lewis, D.: The Voyaging Stars, Secrets of the Pacific Island Navigators. William Collins, Sydney (1978)

22. Koestler, A.: The Sleepwalkers. Hutchinson, London (1959)

23. Thomsen, D.E.: What mean these African stones? Sci News 126(11), 164–170 (1981)

24. Roy, A.E.: The origin of the constellations. Vistas Astron. 27, 171–197 (1984); Roy, A.E.: The lamps of Atlantis. In: Nash, S. (ed.) Science and Intelligence: Proceedings of an IBM Interdisciplinary Conference. Science Reviews Ltd., London (1987)

25. Phillips, G.: Act of God: Tutankhamun, Moses and the Myth of Atlantis. Sidgwick and Jackson, London (1998)
26. Roy, A.E.: 80th Birthday Lecture, Glasgow Science Centre Scottish Power Planetarium, June 2004
27. Vaidya, C.V.: The Mahabharata: A Criticism. Sanskrit Book Department, Delhi (1966)
28. Geil, W.E.: The Great Wall of China. Murray, London (1909)
29. Ralling, C., Heyerdahl, T.: The Kon-Tiki Man. BBC Books, London (1970)
30. Anonymous: Pyramid builder's tomb "found". Daily Telegraph, Dec 2 (2007)
31. Edwards, I.E.S.: The Pyramids of Egypt. Penguin, London (1947)
32. Bauval, R., Gilbert, A.: The Orion Mystery. Heinemann, London (1994); Highfield, R.: BBC re-edits Horizon after watchdog's attack. The Daily Telegraph, Dec 11, 2000
33. Bauval, R., Hancock, G.: Keeper of Genesis: A Quest for the Hidden Legacy of Mankind. Heinemann, London (1996)
34. Lawton, I., Ogilvie-Herald, C.: Giza: The Truth. Virgin, London (1999)
35. Clube, V., Napier, B.: The Cosmic Serpent. Faber, London (1982); Bond, A., Hempsell, M.: A Sumerian Observation of the Köfels' Impact Event. Alcuin Academics, York (2008)
36. Pritchard, J.B. (ed.): Ancient Near Eastern Texts Relating to the Old Testament, 2nd edn, revised and enlarged. Princeton University Press, Princeton (1955)
37. Asimov, I.: The rocks of Damocles. In: Asimov on Astronomy. Macdonald and Jane's, London (1974)
38. Dr. Charles Herberger, F.: Samson strides the skies. Griffith Obs 51(3), 2–13 (1987)
39. Baillie, M.: Exodus to Arthur, Catastrophic Encounters with Comets, revised edition. Batsford, London (2000)
40. Arago, F.: Popular Astronomy. Longman, Brown, Green, Longman and Roberts, London (1858)
41. McBeath, A., Johansson, G.: Letters. In: Cambridge Conference Network, Liverpool John Moore's University, UK, 3 Dec 2001
42. Crommelin, A.C.D.: Diamonds in the Sky. Collins, London and Glasgow (1940), quoting Maunder, E.W.: The Astronomy of the Bible
43. Sagan, C., Agel, J.: The Cosmic Connection: An Extraterrestrial Perspective. Doubleday, New York (1973)
44. Hawkins, G.S.: Beyond Stonehenge. Hutchinson, London (1973)
45. Flightpaths to the Gods. BBC-2 Channel, UK, 21 Aug 1997; Walton, J.: review, Daily Telegraph, 22 Aug 1997

46. Story, R.: The Space-Gods Revealed, a Close Look at the Theories of Erich von Däniken. New English Library, London (1977)

47. Our Science Correspondent: Maya Site is Dated to 2600 B.C. Daily Telegraph, 20 April 1976, quoting University of Bradford Institute of Archeology, University of California and Radiocarbon Dating Research Laboratory, Cambridge, Nature, undated; N. Hammond, The earliest Maya. Sci Am 236(3), 116–133 (March 1977)

48. Dr Krupp, E.C.: Echoes of the Ancient Skies, the Astronomy of Lost Civilizations. Harper and Row, New York (1983)

49. Brown, P.L.: Megaliths and Masterminds. Robert Hale, London (1979)

50. Anonymous: Visitor's Guide to Sunstones II. Lawrence Hall of Science, University of California at Berkeley (1986)

51. MacKie, E.W.: New evidence for a professional priesthood in the European Early Bronze Age. In: Bostwick, T., Bates, B. (eds.): Viewing the Sky through Past and Present Cultures: Selected Papers from the Oxford VII Conference on Archaeoastronomy, pp 343–362. In: Pueblo Grande Museum Anthropological Papers No. 15, Pueblo Grande Museum, Phoenix Parks, and Recreation Department (2006)

52. MacKie, E.W.: Science and Society in Prehistoric Britain, op cit

53. Biswas, S.K. et al. (eds.): Cosmic Perspectives. Cambridge University Press, Cambridge (1989); Lunan, D.: review, Space Policy 7(4), 333–334 (Nov 1991)

4. Archaeoastronomy in the British Isles

It was a black day for British archaeology when I was shown these things.

– Dr. Euan MacKie, describing his introduction to Neolithic
stone rings on Machrie Moor [1].

If the Thoms' and MacKie's work of the 1960s and 1970s is to be believed, for at least 2,000 years, and ending more than 3,000 years ago, the British Isles supported one of the most advanced 'colleges' of astronomers in the ancient world. Their people built stone monuments in France and Spain, and in Scandinavia, and their influence is found in the Mediterranean, leading to arguments about where and how they interacted with other seafaring cultures such as the Minoans. Astronomical alignments were built into their tombs in Ireland, but it's in Brittany, on the English and Scottish mainland, and in the western and northern isles, that their standing stones and circles have the characteristics of observatories.

Almost certainly their interest in the movements of the Sun and Moon was prompted by navigation and by agriculture, and it is hard to imagine that key events such as the solstices would not have been marked by religious rituals. But since no literature has come down to us, and there was no explicit carving on the stones, only the pure astronomy underlying the layout of the sites is still accessible.

Since they did not build in brick, and did not practice metalworking until late in their history, the megalith builders had to reduce positional astronomy to precise observation of horizon events, and these were marked with stones. After surveying more than 500 ancient sites, Alexander Thom published a histogram (Fig. 4.1), showing the measured alignments of the ancient sites in relation to the movements of the Sun, Moon and bright stars, and

D. Lunan, *The Stones and the Stars: Building Scotland's Newest Megalith*,
Astronomers' Universe, DOI 10.1007/978-1-4614-5354-3_4,
© Springer Science+Business Media New York 2013

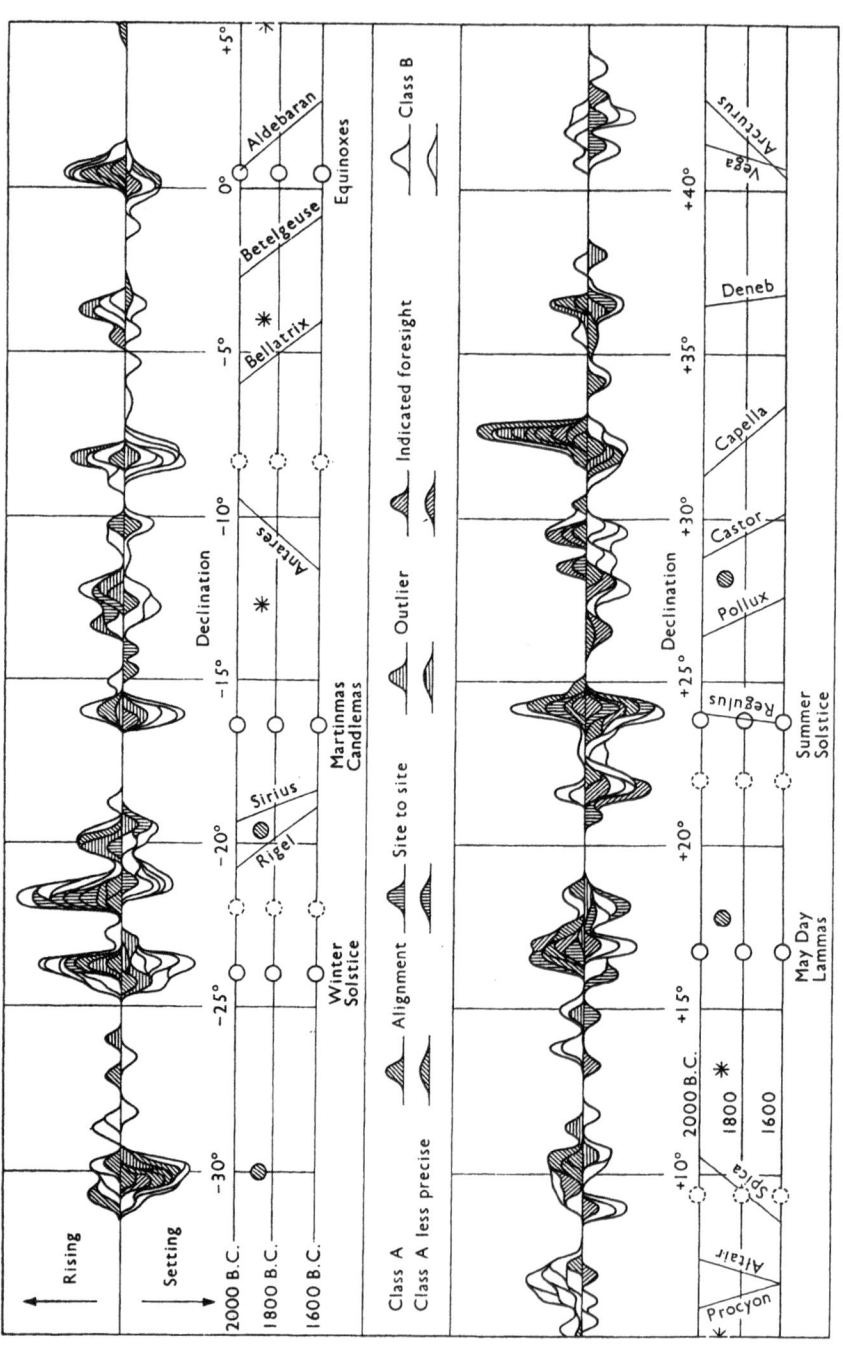

FIG. **4.1** Recurring declinations marked by horizon alignments at Neolithic sites (After Alexander Thom, "Megalithic Sites in Britain [2]")

it was obvious that the relationships were far from coincidental. The recurring intermediate solar positions led him to postulate a pre-Celtic calendar dividing the year into 16 parts [2].

Euan MacKie makes the important point that Thom's histogram was drawn up before the scale of radiocarbon dating was revised, when it was believed that the peak of Neolithic culture fell between 2000 and 1500 B.C. Although that Neolithic period now spans 4000–1800 B.C., MacKie's own research on brochs, the circular stone towers of the Iron Age, has convinced him that the same astronomical principles continued to be applied after the Stone and Bronze ages ended [3].

Still more importantly, the obliquity of Earth's axis was decreasing, so the positions of the solar and lunar alignments in the key to Fig. 4.1 have to be moved to the right to allow for the greater ages of the sites – and the matches for summer and winter solstice, the equinoxes, Martinmas/Candlemas and the other solar calendar dates are improved as a result. So, too, are those for the Moon's four limiting declinations (the major and minor standstills). Extending the inclined lines showing the changing declinations of the stars, due to precession of the equinoxes, gives improved fits for Rigel, Sirius, Antares, Betelgeuse, Procyon, Altair, Spica, Pollux, Castor, Vega and Arcturus, and earlier ones for Pollux, Castor and Capella. The high peak for Capella in 2000–1600 B.C. remains, but there is an explanation for that. Between those dates Capella was becoming circumpolar in the British Isles for the first time, clearing the northern horizon instead of setting, and Thom suggested that the Hill o' Many Stanes at Mid Clyth recorded the observations made as it did so. The revision of the radiocarbon dating scale makes the histogram still more convincing.

One foolish argument, raised all too often in reply, is that because there's no other record of the events, we can't say that the stone alignments point to solstice sunsets, etc., because we don't *know* where those events took place. For all we know, some unknown cause may have altered them, perhaps an undetected shift in Earth's polar axis, or a major variation in its rate of change. It's an argument that might need to be considered if we were talking about just one alignment – midsummer sunrise at Stonehenge, say. But when we're calculating hundreds of them, and they all fit, that could only mean that none of them fitted before and the effect of the supposed change was to bring them all into apparent line. It isn't possible because the positions are actually calculated using spherical triangles in which one apex is the celestial pole, and one side is the co-latitude of the observing site. If the tilt of Earth's axis altered in such a way as to produce an alignment at Ballochroy, say, which wasn't there before, then all of the other sites would be affected differently.

(Some New Age believers maintain that Earth's axis underwent a major shift around 1000 B.C.; if it were true, the whole of megalithic astronomy would be invalid. Their belief in both remains unshakeable. Isaac Asimov pointed out that if this had happened, the consequences would be like the ending of *The Man Who Could Work Miracles*: at the very least, stalactites, thousands of years old and fragile, would have been shattered all over the world [4A]. At one New Age conference a well-known speaker – allowed to remain anonymous here – said, to great applause, that Western civilization's record of violence in the twentieth century would cause the planet's spirit to revolt and destroy us with a polar shift in 2011–2012. Reminded afterwards that earlier in his talk he had mentioned the Aztecs, successors to the Maya, who lived in a state of permanent war to generate enough captives whose living hearts they could tear out in ceremonies they believed essential if the Sun was to rise and the rain continue to fall – "Yes indeed," he replied, "but I have to give the audience what they want to hear.")

In *Megalithic Lunar Observatories*, Thom reinterpreted the Mid Clyth stones, showing that they could also be used for detailed lunar observations [4B]; and even if they were set up to observe Capella, that's not to say that they couldn't be used in other ways. But Thom's lunar interpretations consisted of multiple sites nationwide, working accurately with a fixed standard of length (the 'Megalithic Yard'), moving northward to spread the events further along the horizon and observe the Moon's motion with still greater precision – perhaps realizing that in effect they were discovering the circumference of the spherical Earth.

Many experts found it impossible to believe that such an advanced society could have existed in prehistoric Britain. The picture they painted was of a tribal society where social groupings consisted of a few families at most, and strangers were routinely killed on sight. But even by the late 1970s, circumstantial evidence for an advanced society was beginning to show in other areas; for instance, blanks for Neolithic hand-axes were distributed through-out the country from a very few quarries. In 1977, a Council for British Archaeology symposium at the University of Nottingham revealed that in a study of 3,300 axes, more than 900 (27%) were made of a form of hornstone available only close to Scafell Pike in the Lake District. About 9% were of greenstone found only near Penzance in Cornwall, 9% of a stone from Graig Llwyd in north Wales, and the remainder from just 17 other sites, none of them contributing more than 2% to the whole. Some were moved by sea, from Cornwall to the Thames estuary, for instance, but as many were found within 100 miles of their origin; probably they came overland. "Discussing the question in *Current Archaeology*, one of the editors of the journal, Andrew Selkirk, speculatively visualizes Neolithic youths in Yorkshire, ready for initiation into adulthood, setting off overland for the Lake District on a ritual quest, determined to collect in the shadow of Scafell Pike stone for the axes symbolic of man's estate [5]".

For those found more than a hundred miles from the quarries, especially those in the Thames estuary, the existence of a trad-ing network seems undeniable. Notice that word 'speculatively.' When concerned with shipping axes rather than measuring rods, somehow it failed to provoke accusations of ignorance, madness or similarity to Erich von Däniken.

The evidence suggests that the timing of rituals had primary importance. The burial of the dead, with accompanying ceremo-nial, can be traced to our predecessors, the Neanderthals; and the oldest astronomically aligned structures are tombs. And, when we take into account the experiences of an observer on the surface of the real Earth (Chap. 2), it is not surprising that those first align-ments are solar.

The earliest megalithic tombs – stone-lined communal burial places, roofed over with earth – were created in Portugal and Brit-tany before 4500 B.C. The marked preference for coastal sites strongly suggests that the builders were seafarers, at least along the coast of western Europe; but just who they were the absence of records makes it very hard to say. The revision of the radiocar-bon dating scale has swept away the assumption that the tech-niques of building in stone "must" have come from Egypt, where they began about 2700 B.C., or from Crete or Mycenae more than

FIG. 4.2 Newgrange before the midwinter sunrise, 18 Dec 2009 (Figs. 4.2, 4.3, 4.4, 4.5, 4.6, 4.7, 4.8, 4.9, and 4.10 Courtesy of Boyne Valley Tours, www.newgrange.com)

a thousand years later still. Even the megalithic temples of Malta, whose building began around 3500 B.C., are a thousand years older than the Great Pyramid of Giza.

In Ireland and mainland Britain Neolithic settlement seems to have begun shortly before 4000 B.C., with the building of megalithic tombs starting soon afterwards. The great mound at Knowth was built about 3300 B.C., Newgrange about 3100 B.C., and Tara about 3000 B.C. At Newgrange a carved stone, carefully concealed in the passage wall, is thought to bear the earliest known representation of 'The Man in the Moon' [6] but when the first draft of this chapter was written, Newgrange was known to be special for two other reasons. As far as is known, it was the first megalithic site to have an outer ring of standing stones (Fig. 4.2), and it was aligned so that, at its most southerly rising, the midwinter solstice, the Sun would shine through an opening above the door and right down the long axis of the tomb (Figs. 4.3 and 4.4).

The companion mounds of Knowth and Dowth have similar entrance passages (Fig. 4.5), and they, too, have solstice and equinox alignments (Figs. 4.6, 4.7, and 4.8), while at Tara, the alignment is to the Celtic quarterdays of Samhain and Imbolc (Figs. 4.9 and 4.10) – strong support for Thom's postulated pre-Celtic calendar. One of the carved stones at Knowth may be an actual example of it [3]. It never made sense to say that the alignment of the 'window' at Newgrange was coincidental, but when it seemed to

Fig. 4.3 Newgrange before the midwinter sunrise, 18 Dec 2009

Fig. 4.4 Newgrange before the midwinter sunrise, 18 Dec 2009

Fig. 4.5 Knowth Cairn eastern entrance, autumn equinox 2011

Fig. 4.6 Knowth passage sunbeam, autumn equinox 2011

FIG. 4.7 Cairn T entrance stone and backstone, Loughcrew, County Meath, autumn equinox 2011

FIG. 4.8 Cairn T entrance stone and backstone, Loughcrew, County Meath, autumn equinox 2011

FIG. 4.9 Tara entrance passage, Mound of the Hostages

FIG. 4.10 Tara passage, Samhain and Imbolc quarterdays (Feb 3 and Nov 3)

be unique, it could be dismissed as a piece of theatre, not the product of systematic observation. Even that didn't really stand up to examination; but the multiple examples we now have recall the exasperated comment of the *Apollo* astronaut, when asked yet again if the Moon landings really happened. "If we were going to fake it, why would we fake it *nine times*?"

Finding these alignments in passage graves shows a connection with the theme of death and rebirth, so universally found in the mythologies of agricultural societies. It's sometimes argued that the emphasis on midsummer sunrise is a latter-day, pseudo-druidical romantic notion, while what counted for Neolithic farmers was the northward turning of the Sun after the winter solstice and the guarantee that spring and summer would indeed return. Against that, others argue that the midwinter solstice marked the onset of the worst time of the year, when the weather worsened, the stores ran out, livestock starved or were slaughtered, and cold, hunger and disease took their annual toll of the populace. At this distance the belief structure and the way in which it changed (if at all) are very hard to trace; but it is fascinating to note that very soon after their appearance the apparently astronomical elements took on a "life" of their own – architecturally and, perhaps, metaphorically.

Around 3000 B.C. a large circle was built at Ballynoe, east of the mountains of Mourne. Within it there was a smaller circle enclosing an inner crescent of stones, a small mound, and several stone cists, rather than the huge burial mound of Newgrange. Between 3000 and 2500 B.C. a large stone henge, "the Lios," was erected in Limerick; its precise function is uncertain, but it was not primarily a burial place and may have been a ritual center used also as a lunar observatory. If so, it was one of the earliest.

The passage of the great tomb of Maes Howe in Orkney (Fig. 4.11a, b) was oriented to midwinter sunrise, but the sightlines around it marked other major calendar dates [7]. "Maes Howe's builders were also astronomers," says Neil Oliver in *A History of Ancient Britain* [8]. Its construction was contemporary with the stone village of Skara Brae and was followed in 3100 B.C. by the circle of Stenness (Fig. 4.12), with the highest stones anywhere apart from Stonehenge. Stenness was followed around 2500 B.C. by another great ring at Brodgar or Brogar (Fig. 4.13), where the Thoms believed that accurate lunar alignments were set up outside the ring a thousand years later [9]. The great building projects were now the rings, not the tombs, as if attention had shifted from the dead in the earth to the open sky – but see Chap. 8.

FIG. **4.11** (**a**, **b**) Maes Howe, Orkney, with Prof. Archie Roy at entrance (*right*). Figs. 4.11, 4.12 and 4.13 by Chris O'Kane, 1998

FIG. **4.12** Stenness standing stones (Orkney 1998)

Work at Avebury began in 2900 B.C. with a wooden structure, the first "Sanctuary," which Aubrey Burl saw as a charnel-house for the dead (Chap. 1). Whether that's true, and whether everything else at Avebury took its cue from that, are different questions. What happened next in the immediate vicinity was the creation of the largest manmade mound in Europe, Silbury Hill, between 2850 and 2750 B.C. (Fig. 4.14). There's disagreement between experts about the exact date, as there is about the ditch

FIG. 4.13 (a, b) The Ring of Brodgar (Orkney, sunset, 1998. Vistamorph™)

FIG. 4.14 Silbury Hill (Photo by Linda Lunan, 2010)

and bank at Stonehenge, but it is generally agreed that Silbury Hill and Avebury are contemporary.

There's greater disagreement over how long Silbury Hill took to build. Estimates vary by a factor of 10, from 15 years to 150. However, it is agreed that the stone circles at Avebury were not begun until Silbury Hill was finished, and that Stonehenge II had to wait until Avebury was completed; in terms of commitment of resources (in prehistoric times, manpower), Silbury Hill, Avebury and Stonehenge II can each be compared to the Great Pyramid or, in recent times, the Apollo Project. In *Prehistoric Avebury*, Aubrey Burl views the projects as competition between rival groups, perhaps extending to the enslavement of the Avebury workforce for Stonehenge III [10]! An alternative view is that they were directed by a group with a single purpose, just as, in the space program, Project Mercury and Project Gemini ran concurrently with the early work on Apollo, the three projects producing their results consecutively.

If so, was that purpose astronomical, or was the main incentive still related to the dead? At first glance Silbury Hill might seem to be the greatest burial mound of them all, or the individual burial place of the greatest chief of them all, and it has been tunneled into and excavated on that supposition. But, like the Great Pyramid, Silbury Hill shows no evidence that it was ever used for burial; apparently it was never even planned as a burial place, since there are no counterparts to the unused chambers of the Great Pyramid.

In *The Silbury Treasure*, Michael Dames puts forward an elaborate argument that the hill represents the swollen womb of the Earth Mother, giving birth to the reflection of the Moon in the moat [11]. This interpretation may be overimaginative, but the fact remains that Stenness, Stonehenge I, Silbury Hill and Brodgar were created in short succession and have no overt connection with the dead. Stonehenge I was quite definitely solar – there were only the ditch, the bank and the outlying Heelstone to the left of which the Sun rose at midsummer solstice (of which more later). There were cremations in the ring of 56 "Aubrey Holes" that were dug within the bank, then filled again, but these symbolic acts within the circle are very different from the creation of communal megalithic tombs. The emphasis on the midsummer sunrise, rather than the midwinter one associated with death – at

FIG. 4.15 (a) Avebury from the air, 1980s (Photo by Chris Stanley). (b) Avebury outer ring, with ditch and bank beyond (Photo by author, 2010)

Newgrange, for example, and at the Clava cairns in Scotland – suggests that a new set of interests had developed.

The evidence from Avebury is more ambiguous. There is no immediately obvious astronomical orientation for the structure as a whole or for its outliers (Fig. 4.15); but the "Coves" in the

North Circle and on the Beckhampton Avenue are oriented to moonrise at furthest north and to midwinter sunrise, respectively. As pointed out in Chap. 1, Aubrey Burl's portrayal of ceremonies there is spectacular and quite possibly correct. The dead chieftain may well have been laid in state within the Sanctuary or at Beckhampton Cove until midwinter solstice, symbol of death and rebirth, a propitious time to inter the old chief and anoint his successor. "The King is dead, long live the King…" But the vision of the Moon is wrong.

Going back to the experiences of our "observer on the real Earth" (Chap. 2), let's suppose that he has established his solar calendar and built a circle, perhaps with one or more outlying markers, to commemorate the fact. Over a few years of continued observation he has established that the lunar cycle does not fit evenly into the solar one and, if there is a day-to-day lunar calendar in force, then no doubt some compromise has been arranged, like the assorted lengths of our months or the five "unlucky days" at the end of the Egyptian and Mayan years.

Although Burl said "it was a time when there were no months or weeks," the Maya, the Egyptians and the Romans, from whom we take our modern months, all had lunar calendars before solar ones. Our days of the week are named after sky-gods, and counting by the Moon goes back to the last glacial age and the hunters. By the time the Neolithic farmers reached Britain, they probably had the relationship between the solar year and the traditional lunar calendar worked out, at least in general terms. But sunrise and sunset would need to be watched for the rituals to be timed correctly, and at the major centers permanent markers would be set up.

Our observer, however, is compressing into one lifetime the generations of experience the Neolithic priests had to draw on, so this is all new to him, and he is still watching the Moon and counting days. Whether or not he intended his circle to be an observatory, it will function as one, and draw his attention to the fact that wherever the Sun or Moon rises on the eastern horizon, it will set at the corresponding point in the west. Starting from due east, at the spring equinox, the Sun moves northward day by day until midsummer solstice, then back to due east at the autumnal equinox and onwards to the south until the midwinter solstice. The Moon, by contrast, goes through a similar set of movements in time with its phases, rising north of east for half of the month and south of east for the remainder. But, whereas the Sun comes back to the same most northerly and most southerly points year after year, only next month the Moon either passes or fails to reach the north and south markers set out for it last time.

To the speeded-up observer at the center of the transparent Earth it is obvious that the Moon's orbital plane is slipping backward, pulled by the Sun and also, although not obviously, by Earth's equatorial bulge. The outside observer would find it much harder to work out what was happening, but in due course he would have a full record of the results. Once every 18.61 years the Moon rises and sets at the most northerly point on the horizon it can reach, and, 14 days later, at its furthest south. It rides so high, and then so low, because at the extremes its position north or south of the equator is the sum of the tilt of the Moon's orbit and the tilt of Earth's axis, both measured from Earth's orbital plane. For the next 4.65 years, as month follows month, the most northerly and southerly risings draw together until they meet at due east.

Then they move out again, but only to a lesser maximum which is defined by the difference between the Moon's orbital plane and Earth's axial tilt. Then it's back to due east again, before drawing back out to the most extreme northerly and southerly positions to complete the 18.61-year cycle.

For a while, when the Moon reaches its maximum and secondary maximum positions, the change from month to month will be too slight to be noticeable to the naked eye. For that reason, referring to the appearance of the phenomena to ancient observers, Alexander Thom designated these events as the 'major and minor standstills,' and those terms will be used for the rest of this book. The major standstill, when the Moon is at furthest north and furthest south a fortnight later, recurs every 18.61 years, with the minor standstill 9.3 years later at the midpoint of the cycle. The observer on the ground finds that the markers for solar rising and setting – furthest north at midsummer, furthest south at midwinter – are each flanked by the lunar markers for major and minor standstills. Just to find that out, confirm it, and place the markers with recognizable accuracy, is a considerable triumph for perseverance in an age when the average life expectancy was little more than 30 years. And that brings us to the first point in reply to

Burl's contention that Avebury was not laid out by astronomers. It's not the fact that you could not observe major standstill rise accurately from the Cove that counts: it's the program of observation that was needed before it was built for the Cove to be aligned accurately enough for us to perceive its significance today.

Anyone reading Burl's imaginative reconstruction might be forgiven for supposing that the Moon rises at its furthest north every midwinter. And so it does, but only at its furthest north *for that year*. Its true furthest north, the major standstill, happens only once every 18.61 years. So, if it happens in December in a given year, next time it will be in July, and in the half of the month when the Moon rises in daylight. That's inconvenient for a priest specializing in nocturnal torchlit rituals of death – even if the chieftains are considerate enough to die at 18.61-year intervals – but even more so if the moonrise is in daylight and the phase is small, making the Moon difficult or impossible to see until it's higher in the sky. Worse still, if the condition is that the ceremony be performed at or near the midwinter solstice, as the alignment of the Beckhampton Cove so strongly implies, then the North Circle Cove alignment will be "activated" only 1 year in 37 at best.

The frustrating thing is that the reconstruction is probably correct in most other respects. The alignments of the Clava cairns, Newgrange, and so on, suggest strongly that the midwinter solstice was the time for burials – although the emphasis on midwinter sun*rise* makes the nocturnal ceremony a moot point. Perhaps the proceedings closed with the rising of the Moon? The Moon would rise at some time during the night in at least half of the years, at or near the furthest north for the year. The Cove could still be the center of the ceremony, and its alignment with the Moon's furthest north in its cycle could be symbolic.

Another moment's thought shows that indeed this is what must have been. If the object had been to line up the Cove with actual moonrises then its builders could easily have faced it anywhere between minor standstill and due east, in which case "activation" would have been achieved four times in the 18.61-year cycle. Instead they went for the configuration that could not be observed more than once in a generation and would not be observed at night more than once in an average lifetime. Whatever mystical connotations there may have been, at least one observer and probably two or three generations of them took the trouble to watch the moonrise and line it up. Whatever else it may have symbolized, the alignment of the Cove embodies and immortalizes their effort, and whatever their primary status in the community there is no reason for us not to refer to them descriptively as astronomers. If their main jobs were as shamans, priests, witchdoctors or vets, then perhaps we should call them amateur astronomers, but astronomers beyond doubt – unless, of course, the alignment is a coincidence.

If there were only the North Circle Cove of Avebury to go on, it might be possible to argue that it was coincidence. To paraphrase the Goons, "every Cove has to point somewhere," although, as shown above, after you have laid out a circle the changing positions of the Moon on the horizon should start to impress themselves on you, and lead naturally to an observing program. However, there is a lot more than Avebury to go on; in fact, it was not until the plan of Avebury was finalized and completion was in sight that the action elsewhere really started.

According to Burl's timetable, work on the two circles within the Avebury complex began about 2600 B.C. Work on a third circle was abandoned in 2500 B.C. and instead a great ditch and bank were begun, an earthwork project comparable with Silbury Hill. It, and the stone ring within it, completely enclosing the earlier circles, were completed around 2400 B.C., and work began on the two stone-lined avenues, West Kennet and Beckhampton, which took another century to complete. It was not until then that the bluestones were brought from Wales to Stonehenge, although some authorities think they had been on their way for centuries, and possibly were set up for a time at three sites en route.

At any rate, around 2200 B.C. Stonehenge II was begun as a double ring of bluestones at the center of the bank-and-ditch ring that was already nearly 700 years old. Stonehenge II was half completed when the plan was changed and the bluestones removed, to be replaced by the sarsen lintels and trilithons which we think of as "Stonehenge" today (Fig. 4.16). After several more changes of plan, the bluestones were erected in an inner circle and a horseshoe lined up with the main axis, towards midsummer sunrise. That takes us to about 1550 B.C., and there was still more (nonastronomical) work to be done on the Avenue; but a great deal had been happening elsewhere in the meantime.

FIG. **4.16** Stonehenge today. Vistamorph™ (Photo by Chris O'Kane)

At least 80 stone circles were built in Ireland between 2500 and 2000 B.C. Many of them had the same form as the "recumbent circles" of Aberdeenshire, with stones graded in size away from a giant stone lying on its long side. Burl has demonstrated that at least some of these may have been oriented so that the setting Moon would seem to float along the top of the flat stone at key times. The large stone was always placed in the south/southwest sector of the circles. Such a construction is not particularly accurate or scientific, but it does provide a basis for a growing body of lore on which an observational program could be built.

The evidence for an observing program is found particularly in Argyllshire, in the Western Isles, and in the Orkneys. The further north the observer is located, the more the events on the horizon will be spread out, so that subtle effects become more obvious – once the technique of accurate observation had been developed. To show how that may have happened, we have to go back to our imaginary observer, who had built his stone circle in honor of the Sun or sky-god but came as a result to appreciate the significance of the major and minor lunar standstills.

Such a discovery, painstakingly reached over a lifetime (more probably, several lifetimes) of observation, is well worth recording for posterity. In an age without writing, putting up a stone as a marker is the obvious way to leave a permanent record. For the satisfaction of the individual or group – whether or not sublimated as the ascendancy of the sky-gods over the earlier cult of the dead – the marker should be as precise as possible.

But now the limitations of the circular observatory become apparent. The full Moon, huge as it looks near the horizon, has a diameter of only just over half a degree, which is about a quarter of the width of a knuckle on a fist held at arm's length. (By coincidence, the Moon's slow outward spiral from Earth has brought it to a distance where it looks almost exactly the same size as the Sun, hence the spectacular view we have of the upper atmosphere of the Sun during a solar eclipse.)

To mark moonrise or moonset accurately, to be able to say, "Stand here and the event will be dead on the marker," the marker needs to be more distant than the perimeter of a stone circle – even a great circle like Stenness or Brodgar. The obvious move, then, is to look for natural markers such as peaks or notches on the visible horizon. But, since the sight-lines are now miles or tens of miles long, much greater accuracy can be attained. It becomes possible to work with the edge of the visible Sun or Moon rather than the disc as a whole, and to take a conscious decision as to whether the upper or lower "limb" will produce the most satisfactory result. Over a succession of solstices or lunar standstills, a team of observers with marker posts could refine their observations, automatically averaging out the varying effects of atmospheric refraction along the way, until they found the "perfect" spot from which to see the event. Since with the standstills the precise point might take more than a century to pinpoint, it might indeed seem worthwhile to bring in a great stone to record and perpetuate the achievement. Thom sees the distant horizon-markers as being like the "foresight" on the end of a very long rifle barrel, with the foreground stone as the "backsight" that brings the eye into the correct alignment.

Three crucial questions then arise. Firstly, are there sites where the stones are aligned with horizon features that delineate astronomical events with the kind of precision that ought in theory to be possible with sight-lines of such length? Secondly, are those alignments deliberate or merely coincidental? And, if deliberate, can the work involved legitimately be called astronomical or scientific?

The answer to the first question is an almost unquestioned "yes." At Ballochroy in Kintyre, for example, there are three stones in a line to which the long axis of the central stone is at right angles. Looking along that axis in Neolithic times, one would have seen the midsummer Sun set behind Ben Corra on the island of Jura, with its upper rim remaining just visible over the profile of the hill as it continued to descend. Looking in the southwesterly direction of the line of stones, the upper limb of the midwinter Sun was similarly fitted to the profile of Cara Island. From the observing platform at Kintraw the midwinter Sun would have set completely behind Beinn Shiantaidh on Jura before its upper limb showed again briefly in the gap between that peak and Beinn an Oir (Figs. 1.3 and 5.4). When upright, the great stone in the middle distance pointed like a finger towards the spot to watch on the skyline above it. Lunar alignments are more complex, but examples will be given below.

The critics' case is that the society of the time was not able to interest itself in such matters, so that if the alignments do exist they are insignificant – almost certainly coincidental – or else Thom has imposed his interpretation on the landscape, consciously or unconsciously selecting horizon features that suit his case and then arguing that the stones were deliberately positioned in relation to them. However, Thom worked out how the ancient observers could have plotted their observations on the ground, and at Stonehenge there are postholes, near the Heelstone, in a very similar pattern. Similar grids are found marked out in stone in Caithness, for instance at the Hill o' Many Stanes, above.

However, the most spectacular examples are in Brittany. Even today, thousands of stones are still standing in the great arrays around Carnac on which the Thoms spent so much time. It has been said that, in terms of man-hours, the Thoms' surveys of the Carnac alignments may be the greatest archaeological project of the century; the ground plan is a huge scroll that can be reproduced only in part or by reducing it so much that it becomes virtually incomprehensible.

Precisely how the Carnac alignments were used has still to be established. It's been suggested that they were built by the previous Neolithic hunter-gatherers, representing their defiance as Neolithic farmers took over the territory [8]. But their similarity in type to the Heelstone postholes and the Caithness stone rows is so striking that it can hardly be a coincidence that the greatest of all sites claimed as observatories is in the immediate vicinity. Le Grand Menhir Brisé, the Great Broken Stone of Carnac, was originally 70 ft in length and weighed more than 300 t; it takes us out of the normal size range of standing stones, past the largest stones used in the pyramids, up into the larger sizes of the Easter Island statues. The pieces lie pointing in different directions, as if it broken due to whiplash; in *Megalithic Remains in Britain and Brittany* Thom states: "We agree with Atkinson that the only explanation...is that the lower break, or at least the separation, occurred while the stone was still upright, and that it must have been produced by an earthquake. Experiments with blocks piled on a tray show that it quite impossible to produce the arrangement which we find by any other way than by shaking the tray [9]."

Nevertheless, Glyn Daniel wrote: "Professor Thom insists that the Grand Menhir Brisé was originally upright. There is no proof of this whatever [12]." If the evidence of the break is not enough, then it's hard to see what other evidence there could be. The largest section having uprooted itself as it fell, if it fell, it is now covering the pit in which it originally stood, if it stood (Fig. 4.17a). Even if it never stood, however, that has no bearing on where it was intended to stand, or why. Euan MacKie's excavations at Cultoon on Islay showed that the ellipse there was intended to be an extremely accurate midwinter-solstice marker, but it was abandoned during construction, with stones left lying by their prepared sockets [3]. It's even conceivable that Le Grand Menhir broke during the attempted raising, although that's hard to tie in with the way in which the pieces are lying today.

The important part of the Thoms' contention is that Le Grand Menhir was intended to operate as a "universal foresight" for all eight of the lunar risings and settings at the major and minor standstills (Fig. 4.17b). Finding the site, in their opinion, would have been a task at least as great and as prolonged as the selection, preparation and transportation of the stone, which shares with the great sarsens of Stonehenge the distinction of having been artificially shaped. Small arrays of stones, in the pattern which the Thoms call "extrapolation sectors," survive near two of the apparent "backsights," and it is hard to believe that the great arrays, up among the northerly ones, are not part of the grand designing of a lunar observatory up to 20 miles across.

Glyn Daniel did not share that view. "The stone alignments at Carnac are not locked forcefully to astronomically important sightings. There is no question that many megalithic monuments are

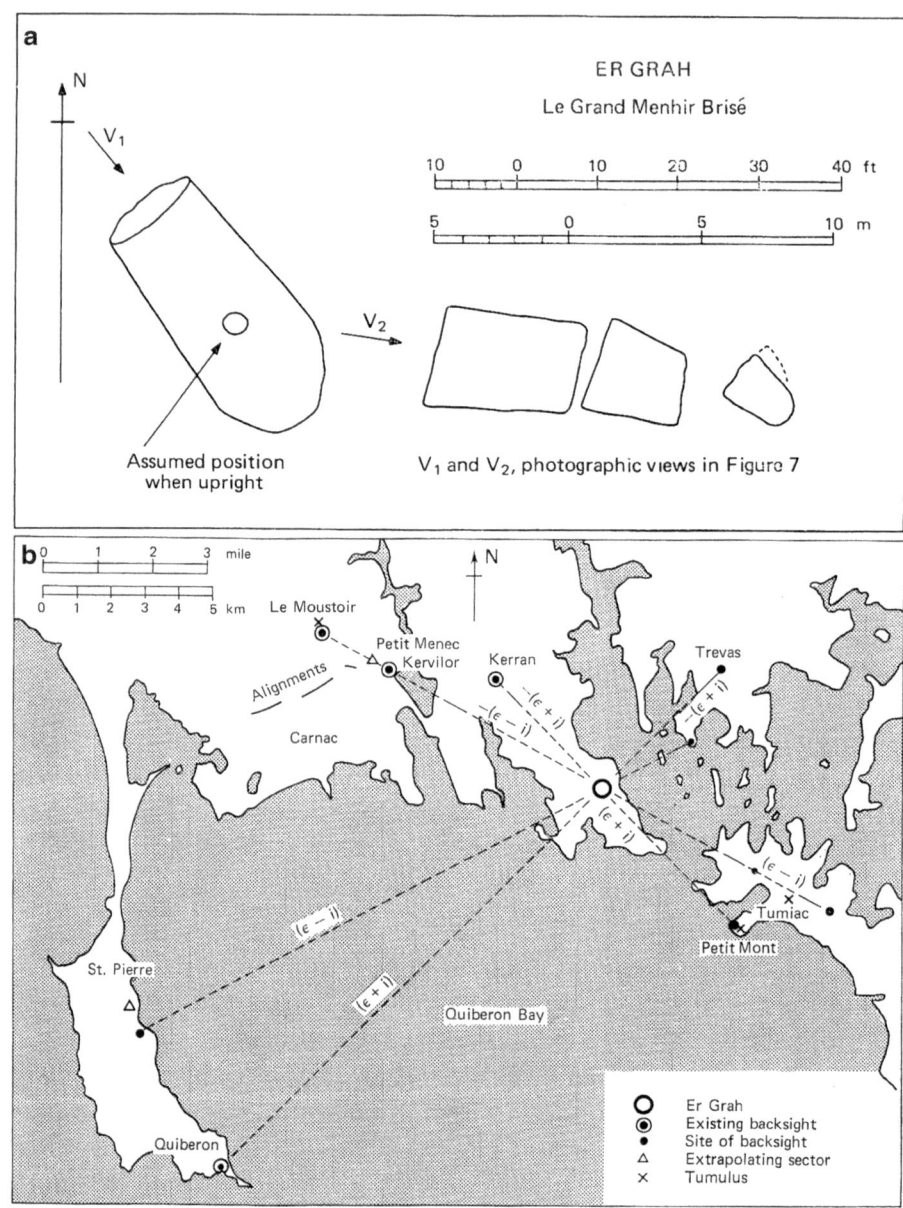

FIG. 4.17 (a, b) Le Grand Menhir Brisé and alignments (After A. and A.S. Thom, *Journal for the History of Astronomy*, 1971)

aligned in a general way, particularly to the east or to the summer sunrise, or to the winter solstice, but that is very different from saying the megalithic builders had astronomical knowledge [12]."

This is what philosophers call a 'persuasive definition.' The Sun is a heavenly body; to align something with its solstitial rising one must know where that is; to acquire that knowledge one must observe, and knowledge of the movement of a heavenly body acquired by observation would be called "astronomical" in any ordinary context. To accept even one lunar-standstill alignment, as Burl does for the Avebury Cove, is to recognize systematic observation over a period of years long enough that some family or group must have been devoted to it. At the very least, they might be called amateur astronomers.

The reasons not to do so turn on the question of motivation – which draws perilously close to a question of semantics. The argument, again, is that the work should not be called astronomical, because it was not done for scientific reasons but for ritual purposes, and the alignments were set up more or less unconsciously as part of Neolithic man's unity with nature; assuming, that is, that they are genuine at all.

Three points arise in response. First, there is little or no evidence for ritual function at the sites that are claimed to be accurately aligned observatories. (Once again, Stonehenge is an exception. The burial of cremated remains in the Aubrey Holes was far more important than early excavators imagined – see Chap. 6.) Second, and more importantly, it can quite reasonably be argued that motivation has nothing to do with the definition of astronomy. The history of the science is full of examples that would be ruled out of court if this kind of hair-splitting were carried throughout. In Chap. 3, we gave the example of scientific work on Project Apollo.

For another, before the invention of the marine chronometer there were determined efforts to find astronomical phenomena that could be predicted with sufficient accuracy for an observer to determine his longitude from the difference between the predicted time of the event and the local time when it was seen to happen. Lunar eclipses fitted the bill, but were too infrequent to be useful. With continuing improvements in astronomical telescopes, however, the eclipses, occultations and transits of the moons of

Jupiter seemed highly promising. But (apart from the difficulty of following Jupiter in a telescope from the deck of a ship at sea) the problem was that the events did not happen as predicted – or, rather, were not *seen* to happen as predicted – because of the time light took to cross the varying distance between Jupiter and Earth. The observations were eventually to lead to the first determination of the velocity of light, by Ole Roemer. It might be argued that the observations and Roemer's deductions from them were not astronomical because they were initially related to the non-scientific practical purposes of navigation. But it would be meaningless to carry the emphasis on "pure" research to such lengths. Likewise, if the megalithic solar and lunar alignments are not coincidental, then they must be recognized as astronomical, on however basic a level.

Thirdly, and most controversially, Thom maintained that the megalith builders pursued their studies of the Moon with a degree of precision that either went beyond the needs of ritual, and was undertaken for its own sake, or else went far enough to allow prediction and deduction, the two great tests of scientific insight into natural phenomena. The first such claim concerns a north–south "wobble" of 9 min of arc superimposed on the mean motion of the Moon. The effect is a small one – a twelfth of a knuckle at arm's length – but translates into a difference in rising position of nearly a third of the Moon's disc, even at the equator. The further north the observer is, the more marked the effect on the Moon's rising position, and the more obvious it would be at a site that had accurately been lined up with the average position of the edge of the disc.

If there were an observing program systematically carried on throughout the British Isles, by a class of astronomers who exchanged their results, then one would expect to find alignments specifically related to the 9 min perturbation, even though such alignments would be "activated" still less often. Obviously it is only at the standstills that we can identify "wobble alignments," and they differ relatively little from the average standstill alignments, so we have to consider how the differences could be marked. One way is to line up a stone or stones with a horizon "foresight," where the Moon's behavior will be significantly different according to the value of the wobble. An example is found

FIG. 4.18 (a, b) Callanish, Isle of Lewis. A view from the sea of moonrise over Callanish inspired Alexander Thom's initial interest in megalithic astronomy (Photos by Mary Goulder)

at Camus a Stacca on the island of Jura, where the Moon setting over Islay at major standstill would just come back into view, with a glimpse of the upper limb, when the perturbation was at its most northerly value. Strong support comes from a site at Knockrome, also on Jura, from which the major-standstill Moon would just graze Crackaig Hill when the wobble had its full southerly value.

The other alternative is to establish two or three nearly parallel alignments bracketing the mean standstill alignment, in relation to closely grouped minor features on the horizon. At Callanish on the island of Lewis, for example (Fig. 4.18), there are two lines of eight stones which relate to the reappearance and final disappearance of the major-standstill (south) Moon behind the hill of Clisham, as varied by the 9 min wobble. At Temple Wood in Argyllshire there are five possible sight-lines to major stand-still (north) setting, three of them marked by stones. One of these aligns with the maximum northerly perturbation, and one of the possible ones would have given the southerly perturbation, plus other subtler variations (Fig. 4.19a). Outside the ring of Brodgar, the Thoms believed, a thousand years after it was first built nine viewing stations were set up, and of these no fewer than seven were "activated" at the extremes of the wobble [9].

Now, unless all the alignments of this class are coincidence (a claim which at this stage seems, frankly, pretty desperate), only two possibilities now arise. Of them Alexander Thom says, in *Megalithic Sites in Britain*: "megalithic Man's interest in the 9 min oscillation probably arose from the fact that eclipses can happen only when the Moon's declination is near the top of one of these waves [9]." Perhaps so, although, as we have seen, the effect could have been discovered just by continuing observation at one of the accurately aligned, more northerly, sites. If the connection with eclipses was *not* recognized, then there are enough sites of this type to suggest that the effect was studied out of pure interest, which sounds like scientific inquiry.

If the connection with eclipses *was* recognized, then the predictions may well have been used to awe the superstitious populace and increase the power of the priests. But inductive reasoning and prediction are usually taken to be the tests of scientific insight, whatever the motivation for the research program. Either way, the northward push of the observing program to sites where the effect could be measured suggests that there was indeed a nationwide class of astronomers.

If the motive was pure research, perhaps pilgrimages were made to the northerly sites as to Jerusalem or Mecca. But, if – as is more probable – the underlying motive was the acquisition of more power for the priests, then it wouldn't do for knowledge of an impending eclipse to be confined to remote sites such as Ballinaby on the west coast of Islay. There would have to be boats and runners and beacons and horn signals to get the word from the islands, if that was where the key sightings were made, all the way to the high priests of Stonehenge and Avebury, if not to Carnac. Given that even in Neolithic times the weather was not

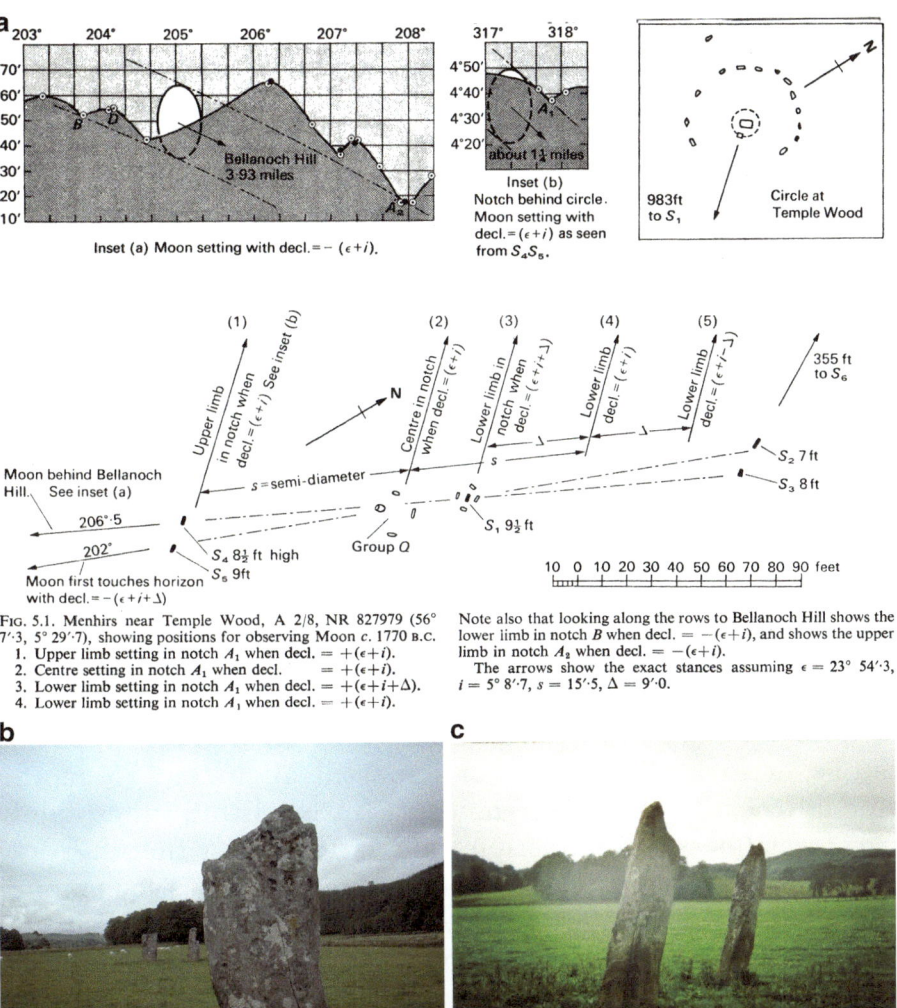

a

203°   204°   205°   206°   207°   208°

Inset (a) Moon setting with decl. = − (ε+i).

317°   318°

Inset (b)
Notch behind circle.
Moon setting with
decl. = (ε+i) as seen
from $S_4 S_5$.

about 1½ miles

983ft
to S_1

Circle at
Temple Wood

(1)   (2)   (3)   (4)   (5)

Upper limb
in notch when
decl. = (ε+i) See inset (b)

Centre in notch
when decl. = (ε+i)

Lower limb
in notch when
decl. = (ε+i+Δ)

Lower limb
decl. = (ε+i)

Lower limb
decl. = (ε+i−Δ)

355 ft
to S_6

N

s = semi-diameter

S_2 7 ft
S_3 8 ft

Moon behind Bellanoch
Hill. See inset (a)

206°·5

202°

Moon first touches horizon
with decl. = −(ε+i+Δ)

S_4 8½ ft high

S_5 9ft

Group Q

S_1 9½ ft

10 0 10 20 30 40 50 60 70 80 90 feet

Fig. 5.1. Menhirs near Temple Wood, A 2/8, NR 827979 (56°
7'·3, 5° 29'·7), showing positions for observing Moon c. 1770 B.C.
1. Upper limb setting in notch A_1 when decl. = +(ε+i).
2. Centre setting in notch A_1 when decl.   = +(ε+i).
3. Lower limb setting in notch A_1 when decl. = +(ε+i+Δ).
4. Lower limb setting in notch A_1 when decl. = +(ε+i).

Note also that looking along the rows to Bellanoch Hill shows the
lower limb in notch B when decl. = −(ε+i), and shows the upper
limb in notch A_2 when decl. = −(ε+i).
    The arrows show the exact stances assuming ε = 23° 54'·3,
i = 5° 8'·7, s = 15'·5, Δ = 9'·0.

b

c

Fig. **4.19** (**a**) Lunar alignments at Temple Wood, Kilmartin, Argyllshire
(After *A. Thom, Megalithic Lunar Observatories*). (**b, c**) Temple Wood
standing stones and major standstill alignment to horizon notch at *right*
(Photos by Chris O'Kane and author)

especially predictable, the reason for having so many sites might
have been to make sure that word did indeed come from one or
more of them. Even in recorded history – in China, for instance –
the penalties for astronomers who *failed* to predict eclipses have
at times been severe.

If, then, we have to accept that the megaliths record a program of scientific observation, by any ordinary definition, how far did the interpretation of those observations go? Thom believed that the builders realized that the world was round and that either it or the sky was very slowly precessing on its axis. As mentioned above, the bright star Capella moved northward spectacularly during the history of the megaliths. It rose and set well south of the major-standstill (northerly) Moon even at the building of Stonehenge I, but it had become circumpolar at the latitude of Mid Clyth by 1760 B.C., never setting at all. All the other stars moved with it, owing to precession, but Capella's behavior, crossing the major-standstill markers, would be most conspicuous, and indeed there are more markers for Capella in Thom's histogram of alignments than for anything else.

As for the world being round, as the builders moved north, and the events on the horizon drew further apart, the astronomers were in effect measuring the curvature of Earth's surface. The significance of it might not have struck them had they not reached Lund, on the extreme north of Unst, the most northerly of the Shetland Islands. There they erected a 13-foot high stone, with a possible minor-standstill alignment to another menhir nearby. At major standstill, in Neolithic times, allowing for parallax and refraction, calculation shows that the Moon itself would have been circumpolar for a day or two each month, for the best part of a year – swinging around the sky without setting. "This information must have been carried south and would have had an effect on these people's philosophical reasoning," Thom wrote in *The Journal for the History of Astronomy*. "Could they have avoided the knowledge that the Earth was a sphere? [13]"

Game, set and match, to Thom? But before returning to the Thom-Daniel debate, which has further ramifications – here is an unprovable speculation from the Glasgow Parks Astronomy Project, formulated by Gavin Roberts and the author. On natural rocks near many megaliths, and sometimes even on the stones themselves, there are carved mysterious "cup-and-ring" marks that some think may have represented stars. Along with them are found carved spirals, frequently in pairs, joined and coiling in opposite directions (Fig. 4.20).

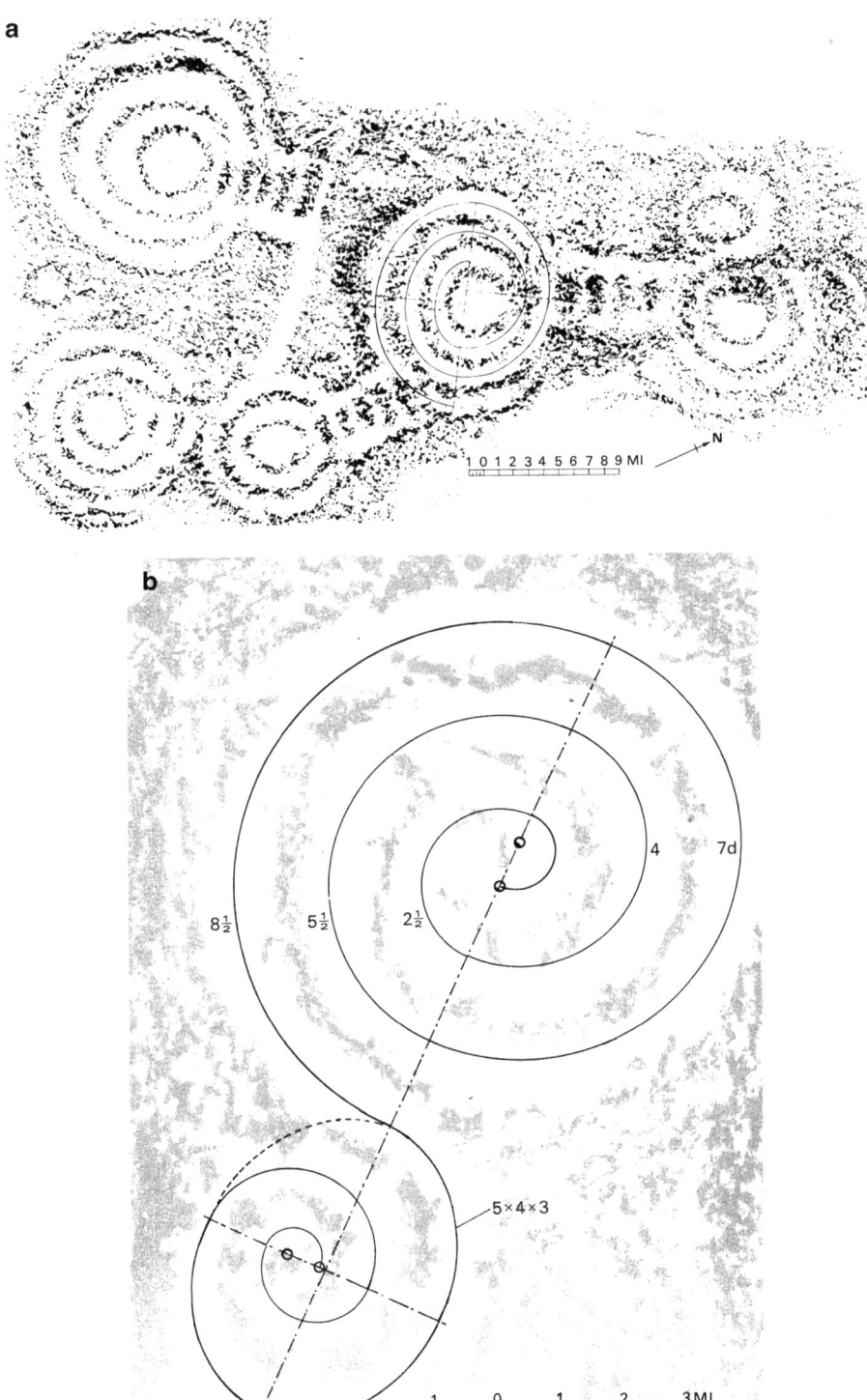

Fig. 4.20 (**a**, **b**) Cup-and-ring marks and spirals (After A. and A.S. Thom, "Megalithic Remains in Britain and Brittany [9]")

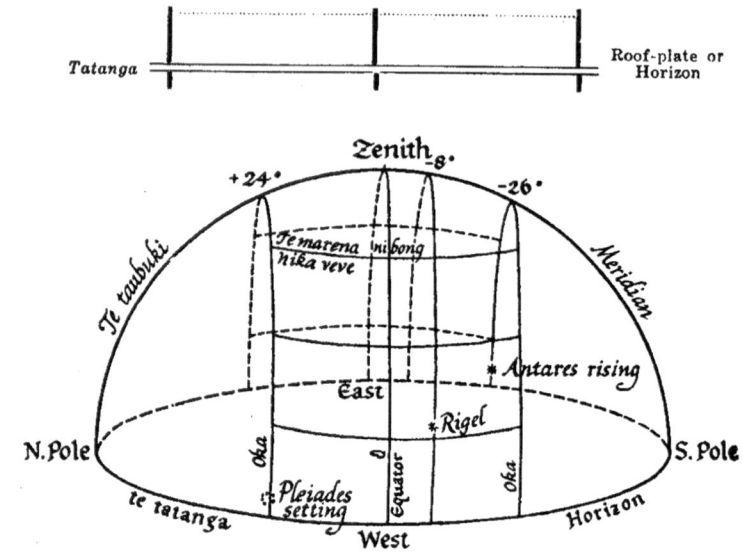

The Gilbertese celestial sphere (after A. Grimble, 1931, and M, Makemson, 1941).

FIG. 4.21 (a, b) Polynesian Star-Ropes (After David Lewis, "The Voyaging Stars [14]")

Now, among equatorial peoples – in Polynesia, for instance – it is quite common for the sky to be divided up not into constellation blocks, as in higher latitudes, but into parallel "ropes" of stars crossing the sky from horizon to horizon (Chap. 3 and Fig. 4.21) [14]. Interestingly, the Mesopotamian myths speak of the sky in this way, as if compiled before some ancient northerly migration. Is it conceivable that the megalith builders' viewpoint was far enough north for them to think of the stars as strung along a great rope, coiling down from the pole to the equator? The figure of Draco, when the pole lay near Thuban as the Avebury circles were built, might have prompted that visualization.

In medieval times, the ascending and descending nodes of the Moon's orbit, around which eclipses occur, were known as the head (Caput) and tail (Cauda) of the Dragon, although the constellation doesn't reach the ecliptic at either end. But perhaps the twin linked spirals, coiling in opposite directions, show a realization that the world was round and that, in the other hemisphere, the "rope" would turn the other way.

b   *The Voyaging Stars*

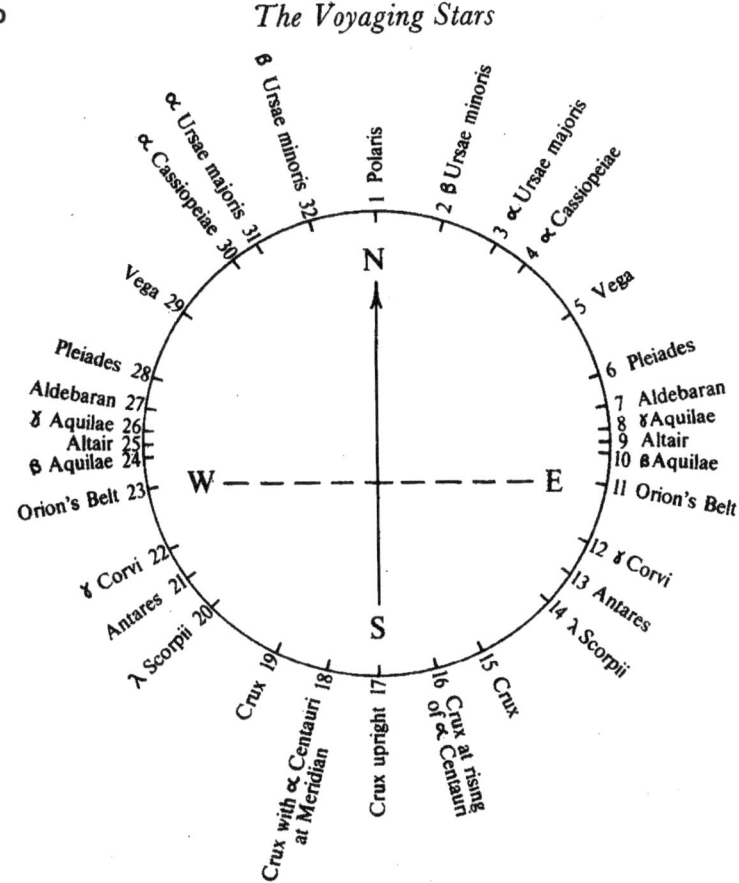

Star compass, Puluwat, Caroline Islands. Positions round the horizon
marked by positions of rise and set of named stars. A similar system
was once in use in the Indian Ocean. (After W. H. Goodenough, 1953.)

FIG. 4.21 (Continued)

If so, it might offer an explanation of the Phaistos Disc, one of
the more intriguing artifacts from the ancient Near East. Found in
1908 at the Minoan Palace of Phaistos in Crete, its context places
it in the second millennium B.C. – if it's not a modern forgery, as is
sometimes alleged. The disc is made of fired clay, about 15 cm in
diameter, and its sides each bear a spiral pattern of repeated symbols, stamped into it – 45 symbols, used 241 times in all (Fig. 4.22),
no more than ten of them resembling hieroglyphs used in Crete
itself. Given the possibility that Minoans reached Stonehenge,

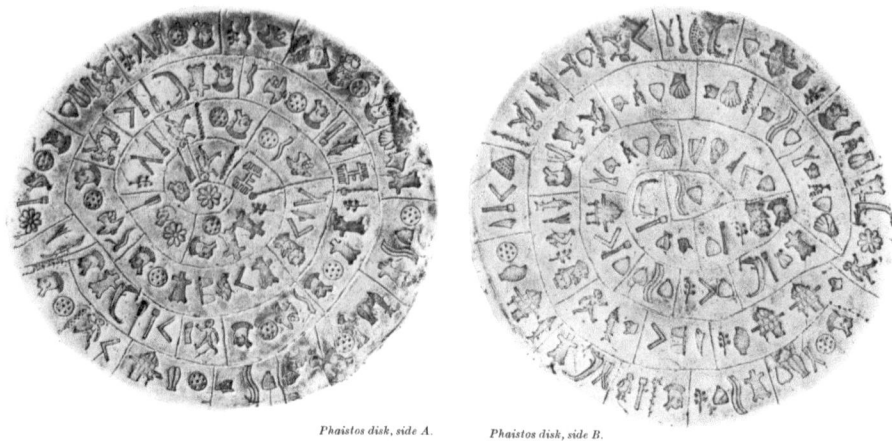

Phaistos disk, side A.   *Phaistos disk, side B.*

FIG. **4.22** The Phaistos disc

evidenced by the carved weapons on the sarsen stones (Chap. 5), is it possible that the Phaistos Disc is an import from the British Isles – a Neolithic planisphere?

Speculation in the field of megalithic archaeology is powerfully tempting, and Glyn Daniel strongly criticized those who engage in it. In his *Scientific American* article he wrote:

> Many people, no doubt bored by the prosaic account of megaliths to be got from archaeological research, jumped on the Hawkins-Thom bandwagon, accepting the builders of megaliths not only as experts in Pythagorean geometry and possessors of accurate units of mensuration but also as skilled astronomers who studied eclipses, the movements of the moon or the positions of the stars. To me this is a kind of refined academic version of astronaut archaeology. The archaeoastronomy buffs, although they very properly eschew wise men from outer space, very improperly insist on the presence in ancient Europe of wise men with an apparently religious passion for astronomy. It seems to me that the case for interpreting megalithic monuments as astronomical observatories has never been proved. The interpretations appear to be subjective and imposed by the observer. Already new surveys are showing the inaccuracy of some of the earlier observations and undermining the hopes of those who believe the builders of megaliths were slaves of an astronomical cult [15].

Leaving the sarcasm aside, the charge of inaccuracy is the nearest this paragraph gets to detailed argument, and Daniel reiterates it in his subsequent letter. Isolated errors have been found in the surveys, although taken together they constitute nothing like enough to invalidate the huge body of the work. Thom's accurate sightlines have been challenged by other surveyors, whose methodology has been challenged in turn by MacKie [3], and his argument will be summarized below. First, however, two independent checks on Thom's approach will be made here.

Thom was a yachtsman, among his many skills, and his interest in ancient astronomy was sparked when at anchor off the island of Lewis, as he watched the Moon rise over the Callanish stones (Fig. 4.18). The sight aroused his engineer's curiosity. Surely the relationship he could see couldn't be coincidence?

Many of his surveys of standing stones in the Hebrides were carried out on his sailing trips, but on the island of Colonsay (Fig. 4.23a), in the inner Hebrides, he was able to make only "a very preliminary survey" because, the night after he arrived, a storm had come up, and his yacht's anchor began to drag. He had to sail off for safety, and the opportunity to return to the island had never arisen. This provided an excellent chance to investigate a set of stones not included in Thom's published surveys, and as part of the Glasgow Parks Astronomy Project, visits were made to Colonsay in 1978 and 1979, with a follow-up in 1980. Even on that one small island, the more stones visited, the more came to light. The Manpower Services Commission agreed to send a survey team over in 1979, but other commitments made that impossible to carry out.

Nevertheless, even rough bearings with compass and clinometer were enough to show that, where the stones had horizon sightlines, some at least were astronomically significant. For example, the standing stone above the hotel at Scalisaig, on the edge of a small circle, appears to have a top shaped to match a notch on the skyline, to a sea horizon, pointing to midwinter sunset (Fig. 4.23b). When that was checked later against the Professor's "preliminary survey," courtesy of Archie Thom, he had identified the same bearing with far greater accuracy.

However, the small circle of stones at the base of the large one suggested a fire site, and that led to finding the possible remains of

Fig. 4.23 (**a**) The island of Colonsay, inner Hebrides (After Frances Walker, 1970s). (**b**) Scalisaig standing stones, island of Colonsay, from upslope. The skyline notch behind contains a sea horizon with an altitude of −1° 30′ (Photo by author, 1976). (**c**) Alignments of some standing stones on Colonsay. (**d**) Key to Fig. 4.23c alignments

two similar sites on a small hilltop and a slope to the east – both of them with prominent horizon notches, and all apparently significant (Fig. 4.23c, d). The accuracy is at least as good as Hawkins's plus-or-minus 2° in *Stonehenge Decoded*, and it's quite probable that a proper survey would show the accuracy found elsewhere by Thom.

b

Fɪɢ. **4.23** (Continued)

Also during the course of the project, contact was made with Tony Crerar, a.k.a. Tony Marchet, who, as well as being a professional mime artist (hence the stage name), spent many years independently finding, cataloging and photographing solar and lunar events on sight-lines and at viewpoints surrounding monoliths near Tyn-y-Cwm in Powys, Wales (Fig. 4.24) [16]. Those were sites never visited by Thom, yet the findings were similar in case after case and backed up by observations. That it took Marchet a big part of a lifetime to make those observations gives an idea of how many people, over how long a time, must have been involved in finding and marking the sites in the first place.

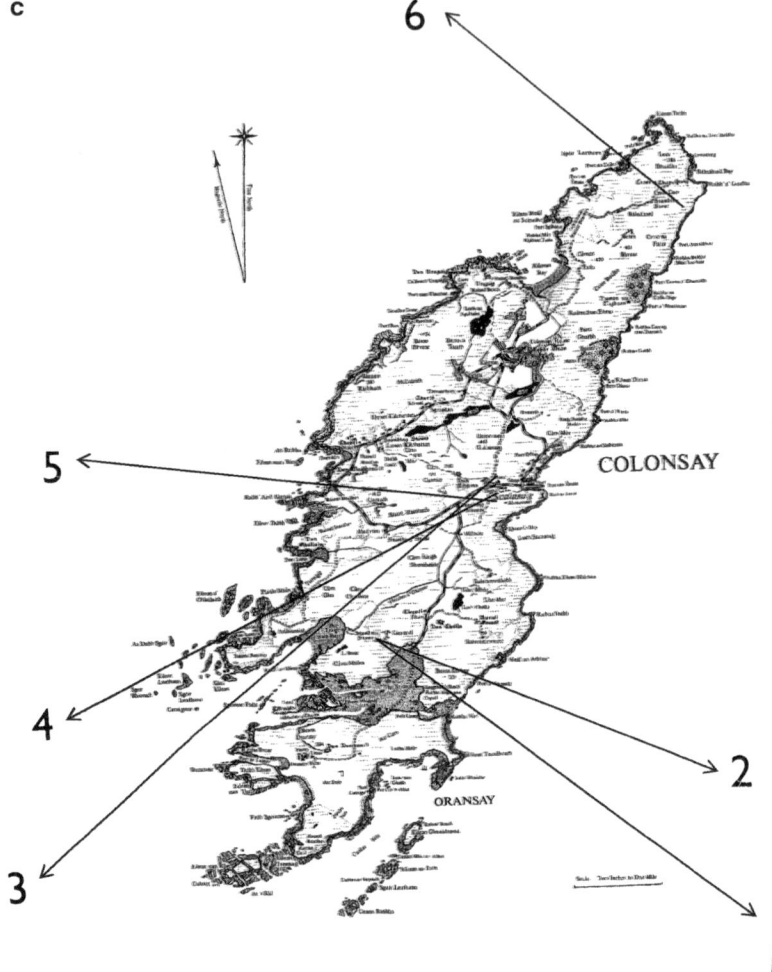

FIG. 4.23 (Continued)

However, MacKie writes

[S]ince the late 1970s Clive Ruggles has systematically and skillfully questioned Thom's astronomical work, starting with a reassessment of 300 of his claimed long celestial alignments on Scottish west coast and in the Western Isles. The conclusion of this critique was that there was no real evidence either for extreme accuracy or for the use of a detailed solar calendar in Neolithic times (broadly the period 4000–2000 B.C.)....

d

As an experiment, the alignments of a number of standing stones to prominent horizon notches were measured in 1979, to see how they tallied with what might be expected if Professor Thom's theories were correct.   There was no prior briefing from Prof. Thom, who had not published his own incomplete survey.   The stones at Garvard (1, 2), Scalisaig (3) and Cnoc a' Charragh (6) were found from Ordnance Survey and local maps plus local verbal directions, and two more possible sites, (4) and (5), by following the lie of the land.   Using compass and clinometer rather than theodolite, in unfamiliar settings, there was no bias to look for already known azimuths.   The results were as follows:

1. Garvard stone (Cnoc Eibriggin, Figs. 9.11A and B) to right-hand notch on Jura: takes in moonrise furthest south, at the major standstill.

2. Garvard stone to left-hand notch on Jura: midwinter sunrise.

3. Scalisaig standing stone (Fig. 4.23A) to sea horizon notch: midwinter sunset.

4. Hilltop, possible fallen stone, to horizon notch: declination approximately −15 degrees, sunset before Martinmas/after Candlemas (cp. A. Thom, Fig. 4.1).

5. Hillside, possible former site of the stone lying downslope, to horizon notch: equinoctial sunset, with early warning of autumnal equinox from sunset notch to right.

6. Cnoc a' Charragh stone to sea horizon notch: declination approximately +8 degrees, sunset on April 10th and September 3rd (cp. Fig. 4.1).

In the valley below stations (4) and (5) above, there are two pairs of small stones (one pair marked on the Ordnance Survey map) which appear to be wayposts; indeed, to naked-eye accuracy, one pair marks the line from the Garvard stone to the Great Cairn overlooking Scalisaig and the ruined village to the north of it.   Yet the compass says its precise orientation is to midwinter sunset, again, while the orientation of the other pair of stones is to the same sunset before Martinmas/after Candlemas, at a different elevation, as if to confirm the alignments of stations (3) and (4).

This study of standing stones on Colonsay is neither rigorous nor complete (for some others, understood or mysterious, see Figs. 4.25 and 9.12); but surely enough has been found to show that Professor Thom's methods do produce meaningful results, even on a first, very rough attempt.

FIG. 4.23 (Continued)

In general, British prehistory has welcomed this authoritative downgrading of Thom's work by someone with similar survey- ing and mathematical skills....

Essentially this author's position is that Ruggles devised a method of resurveying Thom's sites which, although it had the praiseworthy aim of being objective (avoiding identifying standing-stone-indicated horizon markers in astronomically

a

Stone S6, Llananno. Southern Lunar Major Standstill alignment to twin cairns on Drygarn Fawr. circa 1800 B.C.

Declination of the moon:
-29deg.3min.= -(e+i)

This is a synthesized image derived from a survey based on photographic data and computer aided calculation.

Photographs and calculated images:
Tony Crerar tel: 01597 840 616
e: tony@tcrerar.freeserve.co.uk

b

Equinox alignment from Lechwedd Pen Rhiw-wen, Rhayader, Powys.

Photograph of the sun setting at the declination corresponding to the Megalithic Equinox circa 1800 B.C. The semi- diameter sits perfectly in the hill notch to the west.

Photo: Tony Crerar.
tel: 01597 640616
e: tony@tcrerar.freeserve.co.uk

FIG. **4.24** (**a**, **b**) Major lunar standstill and equinox in Powys, c.1800 B.C. (Reconstructions by Tony Crerar)

'suitable' positions) nevertheless also by its nature prevented the identification of accurate long alignments. Thus his conclusion that long alignments did not exist was based on an a priori assumption, not on independent fieldwork and a large part of his subsequent campaign against Thom's conclusions is based on this error...

Drawings A and B on Fig. 5 [of ref. 3] well illustrate how Ruggles's survey method cannot reveal an accurate long alignment even when one exists. Despite the fact that the foresight is the tip of a small island far out to sea, and therefore can mark a sunset precisely, Ruggles's diagram marks with horizontal arrows only the minimum and maximum spans of the horizon indicated by the standing stone itself. Thus a potentially accurate alignment is arbitrarily reduced to a more approximate one [3].

The missing element here is that erecting a standing stone can be regarded as an act of communication, an intention to record or highlight a discovery which the builders have made. The person on the receiving end has to make the 'a priori assumption' that communication is intended and can be understood. (In starting to read this book, the reader has to assume that those conditions are met. If you start with calculating the odds that the ink has fallen randomly on the page in what looks to you like standard English, you'll never get as far as assessing the content.) Recent history provides a perfect example of this point, and why statistics alone can never settle such disputes.

At low tide, the island of Colonsay (see above) is joined to the uninhabited island of Oronsay by The Strand, a stretch of sand dotted with pools of salt water and patches of seriously dangerous quicksand. (Looking at satellite photos of the UK, you can tell if the tide is out because if so, Colonsay and Oronsay appear as a single island. The tide is out so often that one might suspect the same photo-montage is being reprocessed in different ways.) On the Colonsay side, the road slopes down to the sand and stops; the corresponding track on Oronsay is hard to see, especially in rain or mist, so in 1940 a standing stone was erected beside it to mark the safe route across (Fig. 4.25).

The Oronsay shore is relatively featureless, especially in mist, and from the Colonsay side it occupies an arc of approximately

Fɪɢ. **4.25** Standing stone, erected 1940, marking the safe route between Colonsay and Oronsay (Photo by author, 1978)

180°, while the standing stone has an apparent diameter of about half a degree. The odds that it marks the safe crossing are therefore one in 360 – *if you assume that the position of the stone is random*. If you do assume that, and don't drive towards it, you are quite likely to sink. It is in fact better to assume that the stone is there for a purpose; so, too, with Thom's long sight-lines.

In their 'Conclusions' to *Megalithic Remains in Britain and Brittany*, the Thoms wrote, "The main lunar observatories from the point of view of completeness are Brodgar, Stonehenge, Callanish, and Temple Wood. But there are many other places where we find indications of accurate lunar or solar lines. In Megalithic Lunar Observatories we listed 25 under the title 'observing sites.' While some of these may be spurious, it is likely that there are many more scattered throughout the area. Why were there so many? We have seen the midsummer fires burning on the hill-tops in Austria, and in Scotland there were until recently the Beltane fires [and midsummer fairs – see Chap. 5]. These are all probably the remains of a method of synchronizing the calendar in different parts of the Country in Megalithic times. Why then did every small community have its own observatory? [9]"

One reason may have been for recruitment. The Scalisaig standing stone marks only midwinter sunset; but whatever ceremony was performed then, perhaps it was the priest's task to watch for the curious youngster who said, "That's a bit to the left of where it was last year"; included him in the select group watching the calendar stones in the coming year; and if he or maybe she showed real promise, sent the person on to Temple Wood, just as, millennia later, the "lad o' pairts" would be sent to college in Glasgow or St. Andrews. Whether they then returned to their own communities, determined to build still more sophisticated observing sites, or went off to break new ground in communities still benighted, it is easy to imagine them carrying their ceremonial staffs, carefully wrapped against damage, like the young men bringing back their personalized hand-axes at the beginning of this chapter – and that brings us to the geometry of the ancient sites, and the contentious issue of the Megalithic Yard.

Daniel's article went on to offer his own interpretation:

> I see the origin of stone rings this way. First there were circular clearings in the forests that covered Neolithic Europe in the fifth and fourth millenniums B.C. We can postulate that sacred and secular gatherings took place in these clearings. Next, owing to the agency of man's domestic animals and man himself, the forests disappeared, whereupon artificial clearings were created by setting posts in a ring as a stage for similar gatherings. The third phase was the translation of the wood rings into stone rings. Then finally, as the tour de force of a succession of what one can only call cathedral architects, Stonehenge was built in the middle of the third millennium B.C. and flourished as a temple cum meeting place cum stadium for more than 1,000 years. This brings us back to my earlier question: does Stonehenge fit the description of megalithic rings in general as sacred and secular meeting places? The answer seems to be emphatically in the affirmative [15].

Well, yes, but so what? The ceremonial function of Stonehenge is so obvious that it would be pointless to deny it (there will be more to say about that below and in the next chapter). But that has no bearing on the fact that the layout would have allowed accurate lunar observations to be made, and the major-standstill alignment of the Station Stones appears to confirm that they *were* made. At Stonehenge and at Brodgar, the Thoms suggest that the really accurate lunar alignments were not laid out with respect to

Fig. 4.26 Arbor Low (Photo by Chris Stanley)

distant "foresights" until fully a thousand years after the circles were created; what else went on is irrelevant to their argument. However hard one tries not to have "a profound ignorance of the archaeological facts," it's hard to see how it can be anything but "subjective and imposed by the observer" to suggest instead that the circles are symbolic clearings in imaginary forests – forests which, in mainland Britain, vanished centuries before, and which in Orkney and Shetland apparently never existed at all!

The point about mensuration and Pythagorean geometry is, logically, quite separate. The arts involved are not needed for the positioning and layout of megalithic observatories, and whether or not the builders had them is a quite separate argument – a by-product, as far as this book is concerned, of the astronomical one. It arises because comparatively few of the ancient stone rings are actually circles. From the air, Arbor Low, for example, very clearly is not (Fig. 4.26), and neither is Castle Rigg in Cumbria (Fig. 4.27).

Briefly, Thom claimed that his surveys of the internal layouts of the stone rings, spirals and Carnac alignments reveal the existence of a common unit of length, the "megalithic yard" of 2.72 ft, with a standard multiple, the megalithic rod (1 meg. rod = 2.5 MY),

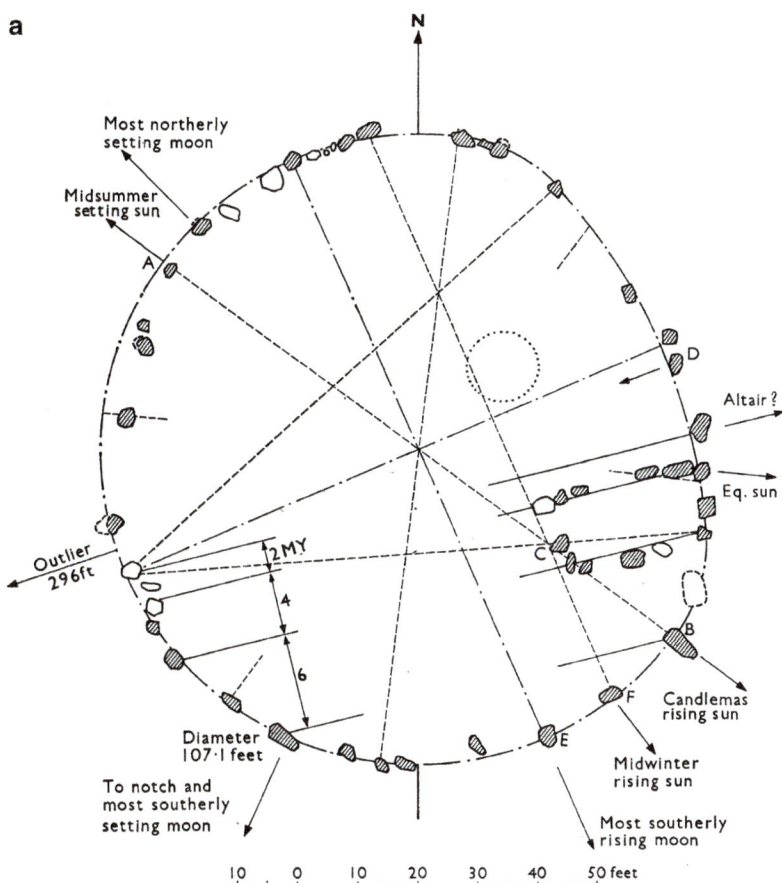

a

N

Most northerly
setting moon

Midsummer
setting sun

A

Outlier
296ft

2MY

4

6

Diameter
107·1 feet

To notch and
most southerly
setting moon

D

Altair?

Eq. sun

C

B

F   Candlemas
rising sun

E

Midwinter
rising sun

Most southerly
rising moon

10   0   10   20   30   40   50 feet

b

FIG. 4.27 (a) Plan of Castle Rigg, Cumbria (After A. Thom). (b, c) Castle Rigg by moonlight, with ghostly figure of DL, during the Leonid meteor shower, Nov 2002 (Photos by Bill Donald)

Fɪɢ. **4.27** (Continued)

and a standard fraction, the megalithic inch (40 meg. in. = 1 MY), used in laying out the spiral carvings. He maintained that the non-circular rings are elegant geometrical constructions of ellipses, "egg-shaped rings," "flattened circles," etc., which were attempts by the builders to make the perimeters of the rings and the radii of the arcs used into whole-number multiple of megalithic rods and yards, respectively (Fig. 4.28). His histogram of measured diameters (Fig. 4.29) shows that half-yards were sometimes used (quarter-yards in the radii). The ground plan often appears to be based on two right-angled triangles, back to back. At Clava, to take a single example, there is a ring that gives perimeter arc radii of 15, 19 and 25 megalithic yards relative to two internal triangles of sides 6, 8 and 10 MY. Referring to these analyses, immediately before the paragraph quoted above, Daniel wrote: "Thom's surveys had already led him to argue for the existence of a megalithic "yard" measuring 2.72 ft and to suggest that the builders of stone rings had a knowledge of Pythagorean geometry 2,000 years before the Greeks. These are extravagant and unconvincing claims: what the builders had was a practical knowledge of laying out right-angled triangles [15]."

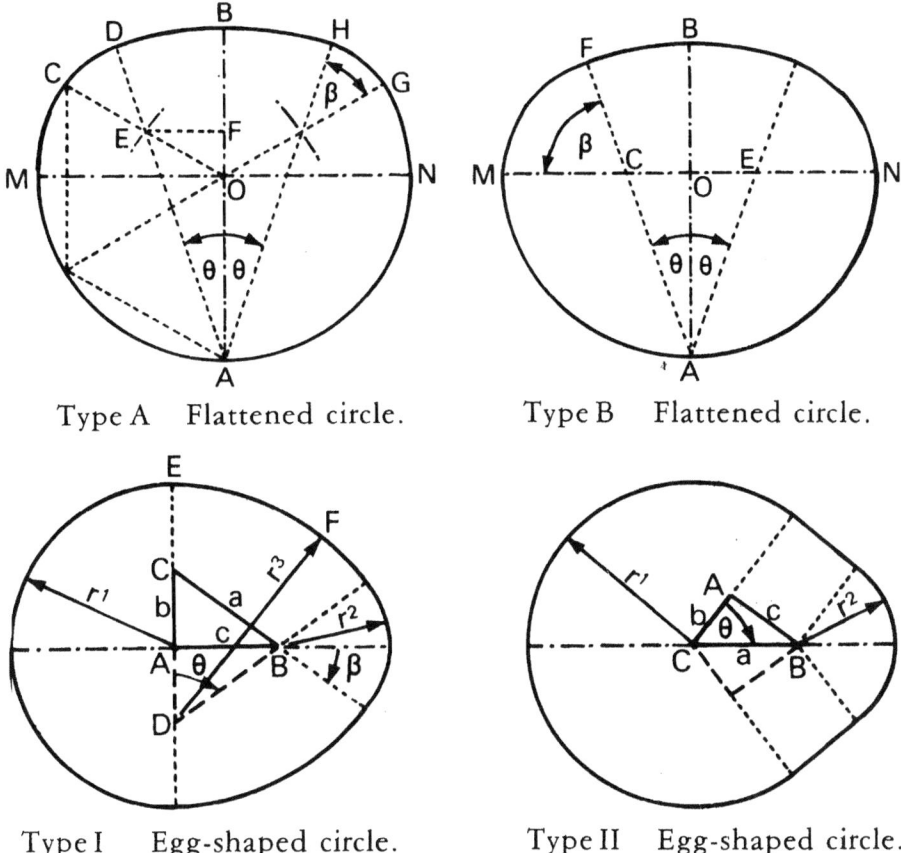

Type A   Flattened circle.   Type B   Flattened circle.

Type I   Egg-shaped circle.   Type II   Egg-shaped circle.

FIG. 4.28 Stone ring geometries (After A. and A.S. Thom, *Megalithic Remains in Britain and Brittany* [9])

The reference to Pythagorean geometry "before the Greeks" is a red herring, intended, like the allusion to "astronaut archaeology" which follows, to attach a discreditable aura to Thom's work. If we took it literally, as the reference to the Greeks appears to imply, then every schoolboy knows what Pythagoras proved about the squares of the sides of a right-angled triangle; but Thom has never, it seems, suggested that the megalith builders were aware of that proof, nor used the squares of the sides in the suggested geometries of the stone rings. In his account, the designers were Pythagorean only in the sense that, like the ancient Greek cult that bore that name, they believed in the mystical significance of numbers.

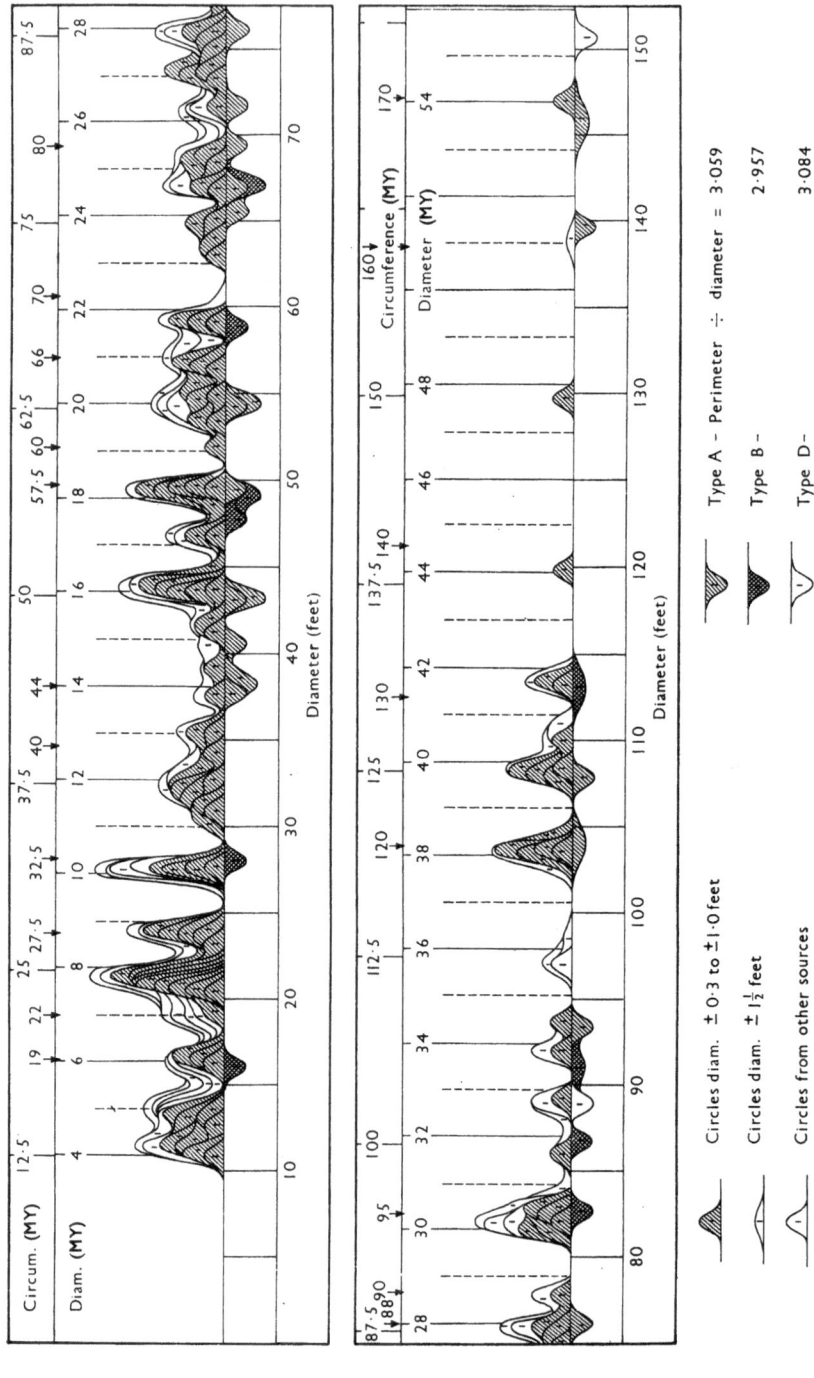

FIG. 4.29 Megalithic yard histogram (After A. Thom, *Megalithic Sites in Britain*)

However, what sells the pass completely is the comment about the builders' practical experience in laying out right-angled triangles. Thom deduced their existence by working back from the positions of the stones to the geometrical principles of their layout, and he didn't suggest that those principles were arrived at by anything other than practical experience. But the experience, in this context, is in the art of laying out large rings in integral multiples of a standard unit of length; in other words, if the layouts are based on right-angled triangles, then the layers-out must have been "into" numerology, and the numbers they manipulated must have been numbers of something – paces, foot-lengths, arm-lengths, lengths of rope or measuring rods.

Thom maintained that the layout of a large number of sites in megalithic yards, up and down the country, was so precise that measuring rods must have been used, and the rods must have been virtually identical, so much so that they must all have been cut to a single standard rather than copied one from another. Comparing the values of the megalithic yard in Brittany and Orkney, for example, the Thoms found discrepancies of only one part in 1,000 [17]. The dispute rested on mathematics, on the one hand, and on the absence of evidence for such an organized society on the other. It was to that apparent lack of absence of evidence that Thom's critics most often referred when they spoke of the archaeological facts – but see below.

When John Braithwaite joined the Astronomy Project in the summer of 1978, he was asked to check out the megalithic yard, because he had a grounding in the relevant math. Having trained in business studies at the University of Strathclyde, he had worked for major companies such as Charles Frank Limited and the defense contractors, Avimo; after the project he went on to found his own company, Dalserf Optics (later Braithwaite Telescopes), which became Scotland's only telescope manufacturer in the 1990s and remained so until his untimely death in 2012.

Braithwaite came back with "a tip of the hat" to Thom's methods and more than a little irked by Thom's critics. One such had apparently maintained that the whole thing was spurious because the measurements disagreed with the megalithic yard ideal by as much as one part in 1,500. Arming himself with the plans, Braithwaite had measured his own modern house and proved without difficulty that by those standards it was not a house, only a random collection of building materials. (There was actually some support for this proposition, when a radiator fell off the living room wall.) John Braithwaite concluded that Thom's sample was large enough for his conclusions to be established beyond reasonable doubt.

Nevertheless, there are recurring doubts. It was said that in at least one case Thom had surveyed a ring and found a good set of values in megalithic yards – although the site was a modern imitation. One has to point out that, in that case, the site may be a *good* imitation, though, as far as is known, the Glasgow project was the first of its kind, and the story may be apocryphal. But, in practice, the megalithic yard is such a universal unit that it is hard to get rid of it. It was decided to avoid using it at Sighthill, in order to distinguish the new circle more clearly from its Neolithic predecessors. Taking the contour of the hilltop as a guide, an arbitrary diameter of 20 modern feet was chosen. But eventually it dawned that the circumference is therefore 62.83 ft, which turns out to be 23.099 MY or 9.24 megalithic rods – not an ideal piece of MY numerology, but closer than one would have liked.

The problem comes to a head at Avebury, where the great size of the structure should provide a definitive test. The Thoms believed that it does, with a perimeter made up of arcs in megalithic rods and adding up to 521 rods in total, all laid out from a right-angled triangle whose sides were 75, 100 and 125 MY. The outcome doesn't seem dramatic enough to be wholly convincing, and Aubrey Burl has quite another explanation for the layout – that the ditch suffered rapid slumping after its completion and before the stones of the outer ring were set in place along it.

Thom estimated that "the two inner circles have a diameter of 125 MY, which curiously is exactly 340 ft." There would be nothing odd about it if the ancient units, like the modern non-metric ones, were based on the average dimensions of the human body. Hawkins suggested, in *Beyond Stonehenge*, that the circles might have been laid out by chains of people so that their differences in size averaged out. Measurements based on the human body are favored also by Professor Burl. The Thoms' reply is that statistical analysis rules out all explanations but the uniform measuring rod; this is just another way of saying that the measurements are very close to multiples of the megalithic yard. Perhaps the measurers-out of circles had all to be of a certain size, like policemen used to be! That is a joke; but it brings us to the ancient country-wide astronomical society postulated by the Thoms and by MacKie, without which the distribution of measuring rods would be not just unlikely but inconceivable.

The objection was that there was no archaeological evidence for a society of this kind. If Euan MacKie's conjectures about the wooden henges, Grooved Ware, etc., were not acceptable, then relatively minor ideas such as the foregoing paragraph were not going to turn the tables. But the curious thing was that, in other areas where there was corresponding evidence, such speculations were apparently acceptable. The report on the highly restricted origin of hand-axes and their wide distribution was a case in point. There was no attack on those researches, although such large-scale regular pilgrimages seemed at odds with a society where strangers were killed on sight, or routinely seized for ritual sacrifice. In a society that was sufficiently organized to permit such journeys, it is no longer ridiculous to imagine novice priests going to the great wooden henges to learn, or, if taught locally, "graduating" on a pilgrimage to make or collect their astronomers' staffs – set to the length of the midsummer noonday shadow of a particularly venerated stone, for example. If the analysis of the axes was valid, it seemed strange to continue to insist that Neolithic society could not have incorporated the megalithic yard.

An early crack in the façade came in 2002 with the discovery of the body nicknamed 'the Amesbury Archer,' or 'the King of Stonehenge.' The man was judged to be between 35 and 45 years old and was buried around 2300 B.C. in a timber chamber 3 miles from Stonehenge, on a sharp bend of the river Avon, with two full sets of archer's equipment, including slate wrist-guards and 15 arrows. He wore gold earrings and was buried with copper knives, flint tools, bone ornaments, boar tusks and other marks of high status – shared with his son, who had grown up nearby and was buried only yards away. He was missing a kneecap, an old injury possibly incurred before he came to England, and that was the most important thing about him, that his bones and his teeth revealed that he had come from central Europe, probably from the area of modern Switzerland.

At a stroke the discovery removed, or should have removed, many of the objections to the Thoms' and MacKie's concepts. Clearly the Archer was highly placed in an organized society, where he was cared for despite his disability, and he had achieved that status despite – or even because of – having come to the area from a great distance. He was buried with no fewer than five examples of a new ceramic ware, fine clay 'Bell Beakers'. After decades of argument about whether the 'Beaker People' were invaders or just the masters of a new technology, we know now that some of them were incomers and were welcomed. It clearly was not a society that consisted of small tribes and family groups, who might come together for festivals or religious construction projects, but otherwise killed strangers and had no spare resources to support a priesthood, far less to heap honors on an archer who could hardly walk.

Nevertheless that image persisted, or took a long time to die. At a temple site on the Dorset Cursus, part of another series of unexplained ancient earthworks, a thousand years earlier between 3400 and 3200 B.C., a 30-year-old woman was buried with three children: a girl aged 10, a boy aged 9, another girl aged 5. The two girls may have been hers, but the boy definitely was not. From the lead isotopes in all of their teeth, it seems that she came from the Mendip Hills, 80 km to the north; she had moved south to Camborne, adopted two of the children, come back to the Mendips to have her own daughter, then come back to Camborne, where she met her end. But was this evidence for a more compassionate, more mobile society? Since the children were buried with her, perhaps they'd all been killed as a punishment for her leaving [18].

However, still more evidence was accumulating for mobility and trade. Great halls, partly used for storage, were found at Balbridie and at Warren Field, in Aberdeenshire, showing the existence of large, stable communities [8]. Part of MacKie's thesis was that flat-bottomed Grooved Ware, decorated by winding a cord around the wet clay pot, was for the use of the priesthood who lived in wooden henges and had furniture (Chap. 1). Now, more than a thousand wattle-and-daub houses have been found with furniture, cupboards, beds and hearths at Durrington Walls (Chap. 5), although the wooden structure near Stonehenge was occupied for less than 50 years [19].

The discoveries of Grooved Ware from Orkney at Stonehenge, and of cattle from across England and Scotland at Durrington Walls, should have caused more of a stir. "Experts therefore believe people must have travelled between the great sacred places of Britain, and that this would hardly have been the experience of the average Stone Age farmer, but the privilege of a new elite, people at the top of society who would have voyaged the length and breadth of the country on a kind of Neolithic Grand Tour…Archaeologists are increasingly of the opinion that the great stone circle of Brodgar and Stonehenge, and the passage graves of Newgrange and Maes Howe, are all memorable phrases within one splendid conversation between the people, the stone, the land and the sky. From the Orkney Islands of Scotland, to the Preseli Mountains in south Wales, to Stonehenge in the south of England and to Brú na Bóinne [the Bend of the Boyne] in the east of Ireland – it is all connected… [8]."

"…as Alexander Thom and Euan MacKie proposed," are surely the words that should follow. But in Neil Oliver's recent *History of Ancient Britain*, from which these quotations come, neither Thom nor MacKie rates an index entry or a listing in the Bibliography. Instead there is an anonymous 'tip of the hat' to Glyn Daniel and his imaginary forest clearings. But the book, the TV series and the subsequent TV 'special' all spell out that if anything, MacKie and the Thoms were too conservative. In Orkney, an entire complex of Neolithic temples has been unearthed (Fig. 4.30), on the Brodgar peninsula, a neck of land between two lochs linking Stenness and the Ring of Brodgar (Fig. 4.31) [8]. Clearly the domain of a resident priesthood for a long period, the oldest level dates to 3300 B.C. but has still to be fully excavated. Eleven pyramidal stone buildings were surrounded by a wall 4–6 m thick and still surviving to a height of 1.7 m.

Around 2600 B.C., the structures were demolished, the remains were buried, and a new pyramid labeled 'Structure Ten' was created, with a cross-shaped inner chamber that is a larger version of the one at Maes Howe. After the erection of the Ring of Brodgar around 2500 B.C., Structure Ten was also demolished, and the nearby stone village of Skara Brae was abandoned [20].

This area is now one of the biggest sites in the British Isles for Neolithic art, with incised decoration, pecked geometric shapes and carved motifs, lozenges, ladders and chevrons. Parts of the

Fig. 4.30 (**a**, **b**) Aerial view of the Ness of Brodgar excavations (Photos by Sigurd Towrie)

inner wall were painted, and an entire 'paint shop' was discovered in 2011 [20]. It turns out that Grooved Ware originated there and was copied elsewhere.

Speaking in Glasgow in September 2011, Neil Oliver invited listeners to consider that Orkney might have been the center of British Neolithic culture, and Stonehenge the group on the periphery, though the pairing of Stonehenge and Durrington Walls could have paralleled Stenness and Brodgar, with the river Avon in the same linking role as the Brodgar peninsula [21].

Fɪɢ. **4.31** The Ness of Brodgar peninsula. Ring of Brodgar on right, Stenness stones to the right of the peninsula, on the far side of the water (Vistamorph™ photo by Chris O'Kane)

FɪɢG. **4.32** Kilmartin Glen (Photo by Chris O'Kane)

And at Kilmartin, that isolated backwater that could never have sustained an advanced lunar observatory, "Something like 350 monuments are clustered within a six-mile radius....It is even thought onlookers might once have gathered on the natural terraces on the valley sides (Fig. 4.32) to watch ceremonies and processions performed among the henges, stone circles and burial cairns on the valley floor below..." Kilmartin was the source of some of the earliest bronze items made in Britain; it was "like a roundabout, a hub where copper from southern Ireland and tin from Cornwall were brought together for onward passage to bronze consumers living in Scotland's north-east..."

Jewelry of Whitby jet was also found in quantity in local graves. "Kilmartin Glen, at the start of the second millennium B.C., may be a snapshot of a moment in time – when Britain found within herself the wherewithal to produce the kind of raw materials and finished products that had currency in the wider world. [Dr. Alison] Sheridan goes further – arguing that by around 2100 B.C. the communities of Kilmartin would have attracted international attention for their metalwork and for prestige items like the jet necklace. She happily describes that part of Scotland, at that time, as 'the epicenter of cool,' as though communities thousands of miles away in the south and east – back in the lands of the Mediterranean and the rest of Europe – would have suddenly become aware that smiths and craftsmen working in the British archipelago were producing items of the highest quality [8]."

And, at a time when all the world was 'watching the skies,' for whatever reason, is it still madness to suggest that with structures that could be used to make observations "of the highest quality," it's probable that they actually did it?

At the time of the Astronomy Project, however, these dramatic revelations lay decades in the future. The setting out of the debate here has been to show that back then, the Sighthill megalith was not built in a spirit of uncritical hero worship but in recognition of the creative work of the ancient builders, and of detective work by the Thoms, MacKie and Roy, into which we looked, to the best of our abilities, before accepting it.

References

1. MacKie, Dr. E.: Prehistoric astronomy: A case study on academic tolerance. Lecture, Glasgow University Department of Physics and Astronomy, and Department of Adult and Continuing Education, Professor Archie E. Roy, 'A Lad o' Pairts', Day of Celebration, 29 Sep 1989
2. Thom, A.: Megalithic Sites in Britain, op cit
3. MacKie, E.W.: Wise men in antiquity? In: Ruggles, C.L.N., Whittle, A.W.R. (eds.): Astronomy and Society in Britain During the Period 4000–1500 B.C. BAR British Series 88, pp. 111–152. British Archaeological Reports, Oxford, England (1981); MacKie, E.W.: The prehistoric

solar calendar: an out-of-fashion idea revisited with new evidence, op cit

4A. Asimov, I., Ages in Confusion. In: Asimov, I.: The Stars in Their Courses, Panther, London, 1975

4B. Thom, A.: Megalithic Lunar Observatories, op cit

5. Getting the axe. Sci. Am., op cit

6. Whitehouse, D.: Prehistoric Moon map unearthed. BBC News Online, 22 April 1999; Whitehouse, D.: The Moon: a Biography. Headline, London (2001)

7. MacKie, E.W.: Maeshowe and the winter solstice: Ceremonial aspects of the Orkney grooved ware. Antiquity 71(272), 338 (1997)

8. Oliver, N.: A History of Ancient Britain. Weidenfeld and Nicolson, London (2011)

9. Thom, A., Thom, A.S.: Megalithic Remains in Britain and Brittany. Oxford University Press, Oxford (1978)

10. Burl, A.: Prehistoric Avebury, op cit

11. Dames, M.: The Silbury Treasure: The Great Goddess Rediscovered. Thames and Hudson, London (1976)

12. Daniel, G.: Letter, Sci. Am., op cit

13. Thom, A., Merritt, R.L.: Some megalithic sites in Shetland. J Hist Astron 9, 54–60 (1978)

14. Lewis, D.: The Voyaging Stars, op cit

15. Daniel, G.: Megalithic Monuments, op cit

16. Crerar, A.R.: The possible significance of the Radnorshire Menhirs, personal communication (Oct 1978); Crerar, A.R.: Advanced lunar astronomy in Bronze Age Wales, personal communication (Mar 1989); Crerar, A.R.: Calculus in prehistoric Scotland and Wales, personal communication, 29 Oct 1997; Crerar, A.R.: Astronomy in prehistoric Mid-Wales, personal communication (Jan 1999)

17. Thom, A., Thom, A.S.: A megalithic observatory in Orkney: The ring of Brodgar and its Cairns. J Hist Astron 4(Part 2, 10), 111–123 (1973)

18. Meet the Ancestor. BBC-2 Channel, UK, 26 Jan 1999

19. Griggs, T.: Riddles of Stonehenge. The Australian, 18 Mar 2012

20. Towrie, S.: Resurrecting the Neolithic priesthood. The Orcadian, 5 Jan 2012; Towrie, S.: The Ness of Brodgar excavations. Islander 2012, 7–12, Orkney Media Group, (2012)

21. Oliver, N.: History of Britain Special: Orkney's Stone Temple. BBC-2, UK, 1 Jan 2012; Lecture, An Evening with Neil Oliver, Waterstone's, Glasgow, 23 Sep 2011

Part II
A Stone Circle for Glasgow

5. Layout and Location

Despite the perks of the visit – the comforts of Shawfield House, the charms of Clementina Walkinshaw, who entertained him there – Prince Charlie was so disgusted with Glasgow that he seriously considered sacking it. The city may have owed its deliverance to the kindly Lochiel who was, as teuchter rebels go, a gentleman. So despite regular but inchoate attempts (every Saturday night) to demolish the fabric, Glasgow was not put to the sack until the twentieth century, when Attila the Urban Planner descended on it like a wolf on the fold.

– The Glasgow Diary, by Donald Saunders, Mostly [1]

The quotation from Edwin Morgan at the beginning of Chap. 3 appears by personal permission of the poet, a former tutor and an old friend, like Archie Roy, Euan MacKie and Donald Saunders, all quoted above. At this point, the story of Neolithic astronomy becomes personal for me. In the early 1960s one of Professor Thom's colleagues and supporters was Dr. (now Professor) Archie Roy, then Senior Lecturer in Astronomy at Glasgow University (Fig. 5.1). He was President of the Scottish Branch of the British Interplanetary Society, and he and I have known each other since the first meeting I attended, in April 1962. In 1963, the Branch became independent as ASTRA, the Association in Scotland to Research into Astronautics, and I started at University and joined Archie's Ordinary Astronomy class. In 1964, he lectured on the astronomical events visible from sites such as the standing stones at Ballochroy, on the west coast of Kintyre, which are lined up with the summits of Cara Island and the peak of Ben Corra on Jura, to mark midsummer and midwinter sunset (Fig. 5.2) [2].

At the next group tutorial the class was set an exercise that began: "You are the astronomer on a colonizing expedition which leaves the Mediterranean in 1800 B.C. and lands in the Mull of

D. Lunan, *The Stones and the Stars: Building Scotland's Newest Megalith*, Astronomers' Universe, DOI 10.1007/978-1-4614-5354-3_5, © Springer Science+Business Media New York 2013

Fig. 5.1 Dr. (now Professor) Archie Roy (*right*), with Howie Firth, Director of Orkney Press, at the launch of "Starfield," edited by author (at rear) (Photo by Chris O'Kane, October 1989)

Galloway. What steps will you take to establish a calendar, given that..." It was a humorous session, punctuated by lines such as "Pass me the chisel: I want to mark the position of Jupiter in *Norton's Star Atlas*" and "Ach, Hamish, it is those astronomers from the Mediterranean again: get out your claymore, man, they are upon us!" But the exercise was to prove valuable 14 years later.

In following years, the interest was such that Dr. Roy organized class visits to the standing stones at Machrie Moor on the island of Arran, where he had discovered the elliptical stone rings (Chap. 4) [3]. For various reasons I was never able to go, though I later visited the site by air (see Chap. 8). In 1965, after more comprehensive lectures by Archie Roy, Dr. (later Prof.) Michael Ovenden and Dr. Robin Green, I visited Stonehenge with my fellow-student

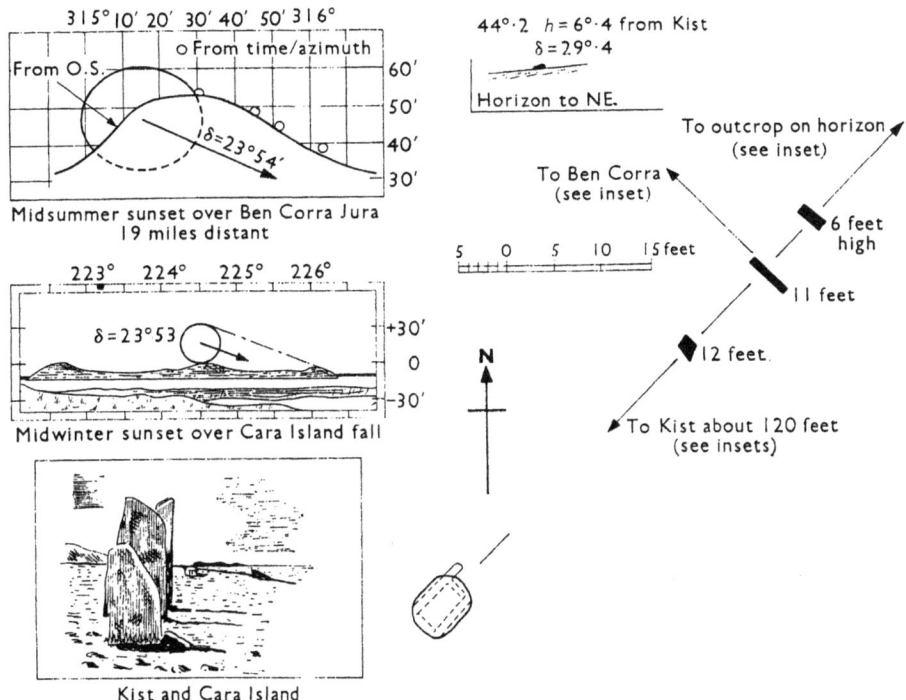

315° 10' 20' 30' 40' 50' 316°

○ From time/azimuth

From O.S.

60'
50'
δ=23°54'   40'
30'

**Midsummer sunset over Ben Corra Jura
19 miles distant**

223°   224°   225°   226°

δ = 23°53

+30'
0
-30'

Midwinter sunset over Cara Island fall

Kist and Cara Island

44°·2 h = 6°·4 from Kist
δ = 29°·4

Horizon to NE.

To outcrop on horizon
(see inset)

To Ben Corra
(see inset)

5   0   5   10   15 feet

6 feet
high

11 feet

N

12 feet

To Kist about 120 feet
(see insets)

FIG. 5.2 Solar events at Ballochroy (after Alexander Thom) featured in Dr. Roy's 1964 lectures

Jerry Bigham from Florida. It was an overcast summer's day; and one interesting point was that from a distance of 20 yards or so we could plainly see the daggers and axes carved on the outer faces of the stones, that according to the guidebooks could be seen only in oblique sunlight in spring and autumn. In form they resemble the weapons of the Minoan and Mycenaean civilizations, and at that time, they were taken to be evidence that Stonehenge has been built by visitors from the Mediterranean c.1500 B.C. [4].

In 1974, following publication of my book *Man and the Stars*, I was contacted by Alan Evans, who was then a captain in British Military Intelligence. He had access to high-resolution photographs of Stonehenge, classified at that time, which showed markings that weren't on the official plans. He believed that they might indicate galactic alignments, which would be staggering if not just an extraordinary coincidence. I checked them out using the dates for Stonehenge that were accepted at the time, and drew a blank. But soon afterwards, an article by Archie Roy in the *Glasgow Herald*

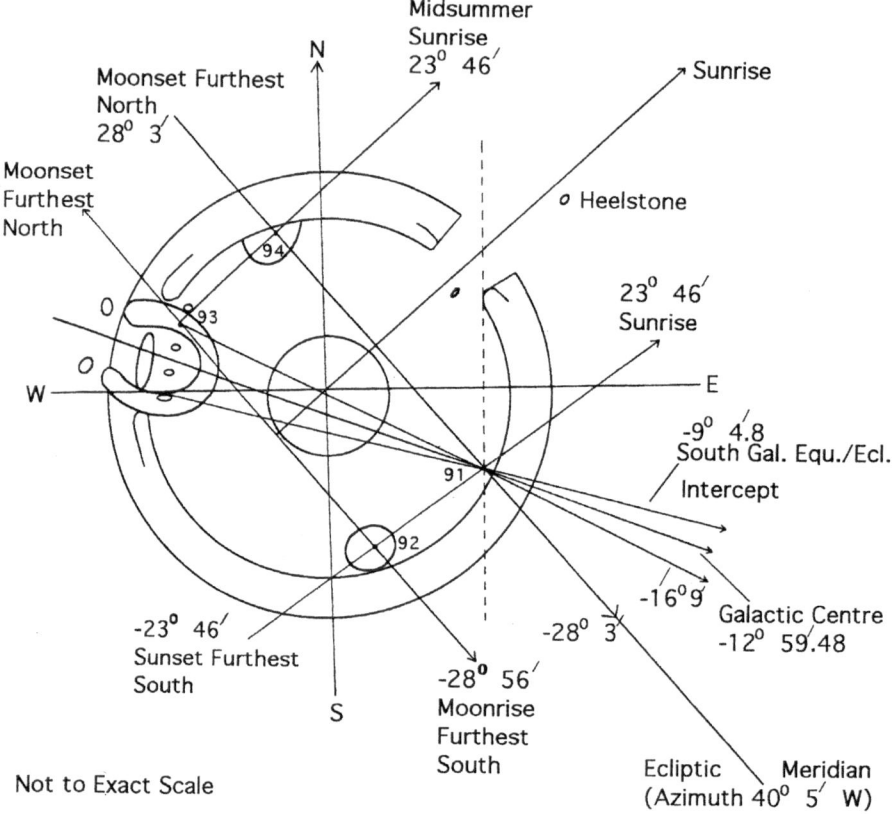

Stonehenge Alignments c. 2840 BC
Without Refraction or Parallax Corrections

Midsummer
Sunrise
23^0 $46'$

N

Moonset Furthest
North
28^0 $3'$

Sunrise

Moonset
Furthest
North

o Heelstone

0

94

93

23^0 $46'$
Sunrise

O

W

E

o
o

-9^0 $4.8'$
South Gal. Equ./Ecl.

91

Intercept

92

$-16^0 9'$

-23^0 $46'$
Sunset Furthest
South

-28^0 $3'$

Galactic Centre
-12^0 $59'.48$

-28^0 $56'$
Moonrise
Furthest
South

S

Not to Exact Scale

Ecliptic
(Azimuth 40^0 $5'$ W)

Meridian

FIG. 5.3 Solar, lunar and galactic alignments at Stonehenge. Drawing by Nick
Portwin, from originals by Alan C. Evans (27 Dec 1975) and author (1977)

revealed that the radiocarbon dating scale had been revised, and
Stonehenge 1 had been redated to 2700 B.C. or earlier [5A]. Rerun-
ning the calculations proved Alan's intuitive guesses to be correct
(Fig. 5.3). I confirmed them optically in the planetaria at Jewel and
Esk College, Glasgow Nautical College, Armagh Planetarium and
more recently Glasgow Science Centre [5B]; nowadays, they can be
verified by any good astronomy program on a PC or laptop.

It might seem they could only be coincidence, because the
position of the galactic center can only be determined using a radio
telescope. The entire galactic coordinate system was reset by the

International Astronomical Union in 1957, after the true galactic center was identified as the radio source Sagittarius A. There would be a partial explanation if the ecliptic was visible at the time. At present, especially in the tropics, a diffuse cone of light can be seen after sunset and before sunrise; in perfect conditions a faint glow can also be detected, on moonless nights, directly opposite the position of the Sun on the other side of Earth. These effects, the zodiacal light and the Gegenschein (counterglow), are both due to the scattering of sunlight by interplanetary dust, confirmed by the *Pioneer 10* space probe in the mid-1970s. Victor Clube and Bill Napier have suggested that they are the aftereffects of a 'super-comet' that broke up among the inner planets around 3000 B.C. (see Chap. 3) [6].

Duncan Steel made the further imaginative suggestion that maybe the zodiacal light joined up with the Gegeschein at that time to form a visible band along the ecliptic [7]. It's just possible to imagine that the light from an explosive event in the galactic center reached us around the same time, so together they could account for the apparent galactic alignments in Fig. 5.3. But what's truly extraordinary is that around 2840 B.C., at the time of Stonehenge 1, the declination of the north galactic pole equaled the latitude of Stonehenge. When the galactic centre rose, the central plane of the Milky Way coincided with the horizon and the galactic pole was overhead. That stretches belief in coincidence to the limit.

In April 1977 I attended the second British Interplanetary Society Interstellar Conference in London and made a short theoretical contribution to an afternoon debate on the Fermi Paradox. I was pressed to suggest specific areas for further research and, with some reluctance, since my direct acquaintance with the field was then small, I mentioned megalithic astronomy. The reaction from the aerospace engineers and the like was amazing. Almost everyone had an opinion on Thom's work, it seemed; most were hostile. Amid a barrage of charges aimed at Thom and Hawkins, one which particularly stood out was that, from the alleged platform at Kintraw (Fig. 1.3), Jura could not even be seen.

That one at least could be checked, and the publication of MacKie's books provided the opportunity. I asked MacKie to ASTRA at Glasgow University Union in March 1978. In the course of the lecture he showed a slide he had taken from the platform, clearly showing Jura in the position claimed. (It appears as a color plate in *The Megalith Builders*, although, to be fair, that book wasn't published until 1977 [8].) In June 1978, on a field trip organized by the Astronomical Society of Glasgow, I had the opportunity to discuss the excavation with Euan MacKie while standing on the exact spot. Although it was too cloudy to see the Jura peaks, the shadow on the clouds showed plainly where they were (Fig. 5.4).

Other controversies were about to interact with the megalithic one. In the later 1970s the Scottish National Party had been making considerable headway with the campaign slogan 'It's Scotland's Oil,' maintaining that Britain's 'offshore' resources of oil and gas should rightfully belong to an independent Scotland. Other opponents of the Labor government were making an issue of rising unemployment, and in 1979 Mrs. Thatcher and the Conservative Party were to fight the general election successfully with the slogan 'Labor Isn't Working.' In an attempt to address both issues the Labor government set up the Jobs Creation Scheme, to be followed by the Special Temporary Employment Program in 1979.

Under pressure from the Trades Unions, Jobs Creation was tied to so many conditions that it became almost impossible to claim the money. The maximum period of employment was to be a year, no permanent jobs were to be created, all projects were to be non-profit, there was to be no competition with any form of unionized labor, and so it went on. Politically, in Scotland, that could be advantageous. It could allow the government to say that when offered an extra share of the oil revenue, the Scots hadn't wanted it after all.

What happened in Glasgow was that, rather like the Norse myth of Baldur and the mistletoe, the planners had forgotten about the Parks Department. For all its industrial reputation, Glasgow, 'the dear green place,' has more parkland per head of population than any other city in Europe; the parks were non-profit, and the employment was seasonal and non-unionized. Everything fitted. Another condition to be met was that the structuring of jobs had to follow a somewhat unusual set of ratios. But the labor-intensive Parks Department was able to comply with those, too. The sum offered to Glasgow was £4 million, and under the formula that was to create exactly 1,001 temporary posts. So the program became 'The 1,001 Project,' with the unofficial slogan, taking off a carpet-cleaner advertisement of the early 60s, "1,001 cleans a big, big city, for only four million quid!"

FIG. 5.4 (a) Standing stone at Kintraw, Argyllshire, from the 'observing platform' on the hillside, where the midwinter Sun set in the notch between the peaks of Jura. A stone directly uphill marks where another observer could give warning of the coming event (Photo by Euan MacKie; detail in MacKie, E., "The Megalith Builders", 1977). (b) The standing stone and notch at Kintraw from the stone platform on the hillside. After dismissing Euan MacKie's theodolite survey as 'a sketch,' a later critic claimed that MacKie's photograph was taken from northwest of this spot [8]. Readers may compare that photograph with mine and reach their own conclusion (Photo by the author, June 1978)

Of course, still more conditions had to be met. The government's planners insisted that "the money mustn't all be spent in one shop." It couldn't all be spent on improving the parks and tending old people's gardens. There had to be Special Projects. This idea was much disliked by the press, with letter-writers constantly declaring that the money should be given to pensioners or to other worthy causes instead. The Parks Department hired an expert from Northern Ireland, Ken Naylor, to come up with suitably imaginative ideas. These included developing methane digesters for the parks' greenhouses, studying Covenanters' graves in the city cemeteries, and many more.

In November 1977 Keith Fraser (Director, Glasgow Parks Department) and Ken Naylor (Assistant Director Landscape) initiated "Astronomy in the Parks" as a Special Project, to be housed with the rest in the offices of the former railway station on Buchanan Street. Knowing nothing about astronomy, as he was the first to admit, Ken Naylor organized a school competition. The winning entry was to build a copy of Callanish, or possibly Stonehenge, for one of the city's parks. As there were no immediate applicants for the manager's post, early in March of 1978 Archie Roy was asked by the Parks Department to take it on. He replied that he was far too busy and, knowing the interest I was taking in the subject, he suggested me for the post. To be precise, he told me on the phone, "I said I knew someone who might be crazy enough..."

At the first meetings between myself and Ken Naylor, the question of a stone circle was scarcely discussed. The brief which had been agreed with the Manpower Services Commission on November 1, 1977, was "the employment of two astronomers to design and erect an astrological construction (sic), such as a mini-Stonehenge using modern materials like steel tubing, as a feature in the Parks. Additionally, the persons employed would be expected to put forward proposals for the promotion of facilities for basic astrological (sic) studies indoors and outdoors, in the City's Parks as an educational and recreational feature with the object of encouraging young people to become interested in astronomy and allied subjects. The project would, it is hoped, develop greater interest in and use of the museums, libraries and the educational facilities offered by the Council and other Institutions in Glasgow."

This grew into a feasibility study for a Glasgow Astronomy and Space Centre, which was discussed at the first meeting in Roy's office, on March 9, and at the official one in the St. Vincent Street Job Centre on March 13, at which I agreed to take the post. Meanwhile, I had brainstormed the Astronomy and Space Centre at the previous Saturday's ASTRA meeting and been to see several proposed sites for it, including the People's Palace museum on Glasgow Green – little knowing the political minefield that this was to turn into later. I still had misgivings about taking the job, knowing that it would disrupt my writing career; indeed, some of the consequences are not fully played out to this day. When Ken Naylor asked if I would take it, I thought hard for a minute and then said "Yes" – only to be told that on paper I had started an hour and a half earlier. It was the turn of the financial year, and if the project hadn't started that morning, the money would have reverted to central funds and been lost.

The Jobs Creation "Astronomy in the Parks" project was to run for eight and a half months, to the end of 1978, and in that time we established that there was indeed a case for constructions such as sundials and a model of the Solar System scaled to the city boundaries. On the educational side there was a considerable demand by schools, libraries and youth groups for talks, etc., but still more for exhibition material to use as the background for class projects and the like. With help from ASTRA we began to meet such requests, and in January 1979 the project was expanded to a much more ambitious "Astronomy and Space Education Program" under the auspices of STEP, the Special Temporary Employment Program.

The program for schools and so forth achieved spectacular results, particularly in the latter half of the year, but by then the writing was on the wall. On taking office the new, Conservative, government had announced that the STEP program would be wound down in favor of YOP, the Youth Opportunities Program, with the emphasis on unskilled school dropouts. I was told that the Astronomy Project, which was at that time fully manned with a staff of ten, had in fact been given by Sir Keith Joseph in the House of Commons as an example of the type of 'nonsense' to be terminated! At any rate, the MSC representative who inspected the Astronomy Project was not interested in what had been accomplished. The only question was whether the skilled adult staff could be replaced by teenagers without educational qualifications. In fact, the Project was terminated in February 1980, by which time I had already left. A great deal of valuable work was wasted, particularly on the construction side, where the sundial to overlook the River Kelvin was waiting only for clearance from the Insurance Department, and the castings for Gavin Roberts's model of the Solar System had been ordered.

Such expanded activities, however, lay far out in the realm of maybes in March 1978. The first month was not easy, and I wasn't encouraged by an article by Brian Aldiss in *Fantasy and Science Fiction* [9], praising me among others for having remained full-time writers despite the odds – just when I had given it up. Ken Naylor had told me I would need to be 'self-motivated,' and as the only team member, that meant that initially I was talking to myself. It took some time to realize that the Astronomy and Space Centre study was not the project's first priority, and that the requirement for a "mini Stonehenge" had to come first. There might be other possibilities, e.g., sundials for the parks, but they would require more skilled personnel and take more time than I had in hand. The Project was expected to close in December 1978.

There was a strong case for building a "modern megalith." As well as fulfilling the MSC brief, it seemed it would arouse public interest and become a tourist attraction. But it would also serve as a tribute by the city to the Thoms, Roy and Mackie, all of whom have close connections with Glasgow. Professor Thom was previously the first Reader in Aeronautics at Glasgow University, and Dr. Thom was Acting Professor of Aerodynamics and Fluid Mechanics there at the time, while MacKie was Assistant Keeper at the University's Hunterian Museum. Roy was Senior Lecturer in the Astronomy Department. By building an astronomical megalith, the first apparently for 3,000 years, the city would be paying tribute to its most prominent researchers in the field. There was no doubt about the degree of public interest, for the growing tourist traffic at Stonehenge had forced the curators to close the circle to the public, to protect the site.

But despite the reference in the brief to a "mini Stonehenge," and the winning competition entry that had proposed a replica of Callanish, the circle could not be a copy of any ancient site. Neither a "mini-Stonehenge" nor a replica of Callanish would work in Glasgow in the twentieth century. The rising and setting azimuths of the heavenly bodies at a particular site are determined by the latitude of the observer as well as by the declinations – which in any case have altered slightly since Neolithic times, due to the change of about half a degree in the tilt of Earth's axis (Chap. 2). Furthermore, star alignments such as the rising of Altair at Callanish have changed totally since 1800 B.C. due to the precession of the equinoxes.

A "mini-Stonehenge" would not work unless the relationships of the outlying Heel Stone and Station Stones were altered, primarily because of the latitude difference between Stonehenge and Glasgow. If the outliers were omitted altogether, then the structure would have no astronomical precision, even if it did include the trilithons and the bluestone horseshoes – for which there would be very little room inside our shrunken sarsen circle. Because of the much greater effect of precession on star positions over the last 4,000 years, if Thom's interpretation of Callanish was used and the alignments were recalculated, then the result would not be recognizable, if it worked at all, even before we started changing solar and lunar alignments to fit Glasgow's latitude and skyline.

The simplest solution would be just to set up a pair of markers in line with midsummer sunrise; it could be done almost anywhere with a view to the northeast, and at very low cost – a couple of concrete posts would suffice. But the Parks Department and the Manpower Services Commission had together agreed to employ four people for a year, and although the start was late, there were still nine and a half months in hand. They had a right to expect something more ambitious, something that would be a public attraction all year round.

Once I had made these points, at Ken's Naylor's suggestion there was a brief flirtation with the idea that a modern monument should commemorate 'invisible astronomy.' Paul Green, the Special Project Manager, had begun looking into it and passed me his copy of Fred Hoyle's Astronomy Today (1975), which he had been reading for inspiration. I made up a list of ten objects, one for each category (black hole, ultraviolet source, etc.), all of which passed over Glasgow, and I began compiling a list of frequencies for a possible radio observatory, but frankly I considered the idea a nonstarter because there would never be anything to see.

As I saw it, there were three different and interacting requirements to consider. The first was that there had to be something for the day to day park visitors to see; the second was that the structure should be a tribute to or at least a recognition of the ancient astronomers and their modern interpreters; and thirdly, therefore, there had to be reasonable astronomical accuracy. Roy pointed out the problems involved. Within the confines of the circle of any practical size, it would be impossible to indicate any alignment with better precision than several solar or lunar diameters. Even if I used very thin markers, their parallax relative to the horizon would be significantly affected if the observer shifted his weight from one foot to the other, or even shut one eye.

Outliers to be seen from the center of the circle, like the Stonehenge Heelstone, would be an improvement but still only indicate where to look, as the stone at Kintraw points to the notch in the profile of Jura. The problem would be to find somewhere in the city with a clear sight-line to a distant "foresight" and vacant ground on which to build a megalith. Variations in atmospheric refraction, within a modern city, would complicate the problem (as indeed they did – see later). Extended observations would be needed, furthermore, to make sure of the alignment; and the chances of having more than one natural foresight would be remote.

Archie's suggestion was that I should compromise on accuracy and accept the limitations of working within the city. The view from his office was marred by a relatively new factory chimney down by the Clyde, in the approximate direction of midwinter sunset. From a distance of a mile or more, it could provide a reasonably precise marker for the last gleam of sunset, perhaps with the upper limb just peeping out after passing behind the chimney, before finally vanishing. The exact viewpoint would be critically dependent on the altitude of the horizon, the height of the viewpoint above sea level, and using a middle distance viewpoint there would be lots of refraction to contend with, so the viewpoint would have to be determined at the solstice by a team of observers in the traditional manner. The time to midwinter would be best spent in evaluating promising viewpoints, in relation to open ground available.

Again, however, there would just be a single alignment and a structure, circular or otherwise, would be redundant. The work that was put in would be a repetitive exploration of the city for spires, chimneys or tower blocks on the horizon from viewpoints that lay along the midwinter sunrise or sunset bearings. There would be nothing to show for it, and the only action would be compressed into a few minutes at the solstice, which could all too easily be rained out. I would need help only for those few minutes, furthermore, and I did have some obligations to generate a proposal that would create temporary employment – although so far there had been no applicants for the vacancies.

We could set ourselves the task of designing a structure around a single alignment, perhaps a recumbent circle or an ellipse, but since we weren't planning to hold ceremonies it would be non-functional and intellectually false. Lastly, if the alignment was related to the skyline of the city, then new buildings or demolition could totally invalidate it, and Glasgow has seen several major shifts in planning emphasis over the last couple of decades. Almost anything apart from the Cathedral was liable to vanish without warning. To be a real tribute, our alignments should be determined by the natural skyline and intended to last for as long as the prehistoric megaliths themselves, to say something to the future about our knowledge of them, regardless of modern buildings flickering in and out of the way for a century or so.

On still another hand, people visiting a park weren't liable to be interested in where the Sun would rise "if it wasn't for the ´ouses in between." Taking that principle to absurd lengths, one might as well build a megalith in a traditional back green, up a "close," completely hemmed in by tenements. Indeed, it turned out that the University of Strathclyde had already done just that with "Steelhenge," a modern sculpture inspired by Stonehenge, but sited in a quadrangle almost entirely surrounded by buildings (Fig. 5.5). It might well be necessary to limit an operational megalith to just one solar alignment, or quadrant if there was room for the flanking lunar standstills.

One of my first steps was to calculate the theoretical alignments for rising and setting at the solstices and lunar standstills. Initially I did a basic calculation of the rising and setting azimuths of the midsummer and midwinter Sun on the true horizon at the latitude of Glasgow, plus the flanking alignments of the lunar 'standstills' generated by the 18.61-year revolution of the nodes of

Fig. 5.5 'Steelhenge' on the campus of Strathclyde University, by Gerald Laing, 1974 (Photo by Linda Lunan, June 2010)

the lunar orbit, marking out four arcs of the horizon. Corrections for parallax would have to be applied to the lunar alignments, of course, because we were on Earth's surface instead of at its center. The effect of parallax is that a heavenly body appears lower in the sky than it really is, relative to the center of Earth, and the displacement has its maximum value when the body is on the horizon. The stars are so far away that even their horizontal parallax is negligible; for the Sun the average figure is only 8.8 s of arc, too small to affect my efforts. But for the Moon the value is a full degree and could not be ignored.

I could neglect it meantime, however, because the effect of refraction is to bend light downwards as it enters the atmosphere, thus making the heavenly body look higher in the sky than it otherwise would. Furthermore, both parallax and refraction have maximum effect at the horizon and diminish towards the zenith, so there would be a ticklish job of multiple calculations to do once the actual height of the horizon had been determined, relative to the true horizon at the site chosen. The effect of parallax might be canceled out or even reversed; and rather than halt the calculation in mid-refinement and risk confusion later, it would be better (and adequate) to work with purely theoretical figures, as if Earth had no dimensions and no atmosphere (the second observer's viewpoint in Chap. 2).

At first I intended to include the rising points of Sirius, now and in 1800 B.C. at the end of the megalith-building era. As the brightest star in the sky, it was most likely to show against city lights, and the contemporary alignment would pinpoint the date of the structure for hypothetical future researchers, if the origin of the circle had been forgotten. If some latter-day Gerald Hawkins ran the alignments through a computer, both dates would come up for the same star, and excavation should reveal which was correct. It turned out, however, that over the last 3,800 years precession of the equinoxes has moved Sirius almost parallel to the celestial equator, so its rising point had scarcely altered. I substituted Rigel, at the foot of Orion (see Chap. 8), which has moved significantly (Figs. 5.6 and 8.27), and is also the star second most often pointed out by markers at the ancient sites (Fig. 4.1).

At this stage, however, there was a hiatus. Nothing more could be done in this particular direction until I had a site, or at least a choice of sites, for which I could produce designs appropriate to the groups of events visible from them. But there was nobody available to tell me how to proceed. It was the end of the financial year, and the senior people in the Parks Department were simply too busy to see me. Just looking at the map was no way to tell what parts of the skyline could be seen from where, and with one of the largest totals of parkland area among British cities, random physical exploration would be a huge task.

Salvation arrived one lunchtime in the shape of Tom Bradley, one of the Parks personnel assigned to liaison with the Manpower Services Commission. Tom's job was to make things happen – to make sure that the innocents like me did get the information and practical help that they needed from within the formal structure of the Parks Department and more generally the District Council. Since project heads like me were expected to be around for only a year or less, it was inevitable that much guidance would be needed.

For my unusual requirements, the appropriate authority was Ronnie Gray, the Principal Landscape Architect, whom I met on

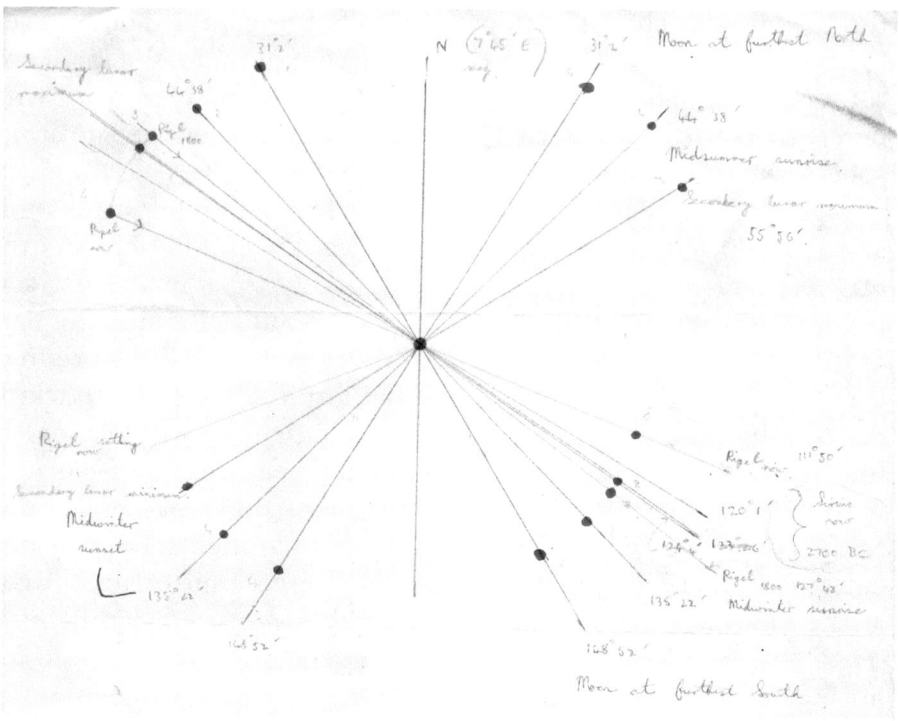

Fig. 5.6 Rain-soaked pencil drawing used with the Miner's Dial, April 1978

April 27. It was clear that he had mental pictures of the skyline of every park in the city, and once I had explained my objectives, he went to work on a large photocopied map to mark the possibilities. There were more of them than one might suppose. As the Lord Provost is quick to point out in relation to the proposed devolution of government agencies, or potential industrial development, Glasgow is one of the few cities whose design allows one to reach open country by car in 20 min or less, in almost any direction, and *you can see out of it.*

The city has kept to the river valley, not encroaching on the natural skyline of the Campsie Fells in the north, nor on the Southern Boundary Fault, yet it is itself on the lower hills of glacial drumlins. The Jobs Creation projects were housed in what had been the booking office of Buchanan Street station, once the city's rail terminus for the north, and only yards away from the junction with Bath Street – about as central as one can get. But the line of sight down Buchanan Street went clean over the south side of the city to Cathkin Braes. Most of the old parks were centered on hills with similar views; the only problem was liable to be that the Victorians had a great fondness for planting trees on the summit of the parks, so that the views were either limited ones from the surrounding slopes or restricted to avenues.

Glasgow Green, which I had already investigated as the oldest park (1662) and because of its astronomical associations, was just across the river from the Nautical College, which at that time had two telescopes and a planetarium that was shown to visiting parties. Glasgow Green surrounds the People's Palace, which for many years was the home of the Fulton Orrery – an enormous creation demonstrating the Solar System out to Neptune. At the time of the project the orrery was in storage, and after some years at the Museum of Transport, it has now been reassembled at the Kelvingrove Art Gallery and Museum. The main attraction from the megalith viewpoint was an imitation Egyptian obelisk commemorating Nelson, but unfortunately Glasgow Green is low-lying and flat, so the skyline is filled by trees and surrounding buildings. The obelisk isn't aligned to the cardinal points, either.

In all, 18 possible sites were marked for investigation for me. To make the initial checks, I was entrusted with a Miner's Dial – a giant compass on a heavy duty tripod, beautifully made in brass and kept in a padded wooden box. This device was so useful that it was to reside with the project, off and on, for more than a year. Its one peculiarity was that it showed west to the right of the north–south line, and east to the left, with 360° bearings likewise running round the dial the "wrong" way. The explanation is that because mine galleries are often awash – especially in rescue situations where maps are most needed – mine maps are drawn to be read by the helmet lamp of a man lying on his back, holding the map up to the roof. The Miner's Dial was designed to be used aboveground in conjunction with the below-ground maps.

If we had used the Dial upside down, it would have matched the orientation of the sky. In 1973 that led to considerable confusion at a lecture I gave to the British Interplanetary Society, mentioning the southwest proper motion of Arcturus. This was challenged by Alan Bond, who said it was southeast. My source had been Camille Flammarion's *Les Étoiles*, and Bond later agreed that the direction of Proper Motion was shown correctly to the right and down, as seen by a northern hemisphere observer. We supposed that there was a printing error in *Les Étoiles* and a correction duly appeared in the July 1973 issue of *Spaceflight*. Unfortunately Flammarion was right the first time: the direction is southwest because it is towards the western horizon.

At the time the Miner's Dial was made it would also have been correct for the situation on the Moon. Historically Moon maps were drawn with south at the top, as it appears in a refracting telescope from Earth's northern hemisphere. But to tally with star maps, they showed west on the left and east on the right. Thus on the Moon the Sun rose in the west, though the Moon rotates in the same direction as Earth does. At the time of Project Apollo the practice was discontinued because it might lead to dangerous confusion. But the Miner's Dial wouldn't work on the Moon anyway, since the Moon has no intrinsic magnetic field.

If Moon maps, Miner's Dials, etc., survive into the future, origins forgotten, a latter-day Velikovsky or von Däniken may argue that Earth and the Moon "must" have turned the other way. If some latter-day Thom from an engineering background tries to deduce their true function, what will archaeological critics say about him?

Tom Bradley arranged an official car and driver (it was to be the last time the astronomy project attained that prestige!). The weather favored us with high winds and lashing rain, and although Tom was keen to see how the search turned out, he took his leave at lunchtime in search of warmth and dry clothes. This was a pity because the first site visited in the afternoon was to prove the winner and looked it from the start, although in those dreadful conditions I could only guess at the distant skylines. But I could at least assess the sites in term of near and middle distance views and obstructions. In fact nearly all 18 spots had some astronomical possibilities. Just in case anyone ever considers building another megalith for Glasgow, here is the full list with notes I made at the time (April 28, 1978):

1. Broomhill Park (now Sighthill Park). Several good sites, with all-around skylines.
2. Alexandra Park. Summer sunrise, winter sunset only; both framed by trees.

3. Easterhouse. Waste ground [then]; future possibility. Skyline quite good for 360°. Penston Road/Londay Road good skyline except N-W, and summer sunrise behind pylons.

4. Tollcross Park. Good skyline from mound overlooking football area. Houses close due east.

5. Queen's Park. Like Kelvingrove: most events can be seen but not from same point.

6. Castlemilk. Good summer, N. skyline; winter rise behind trees; winter set behind flats. Leave site for kids.

7. Cathkin Braes Park. Summer sunrise through trees, winter clear; summer and winter sunset obscured. Better horizon higher on hill, but nearer power station. Better spots further west on Cathkin Road.

8. Glenmuir Drive. Area hopeless.

9. Priesthill Road. No chance.

10. Bellahouston Park. By Exhibition monument all solar, lunar alignments clear, but between trees on hill, by existing monument. Sirius rising no good.

11. Kelvingrove Park. Views of Midsummer sunrise and set, also SW and SE, but not from same spot.

12. Ruchill Park. View of sorts to NW, SE; good to SW, trees to NE. Consider Hillend Road?

13. Knowetap Street. Midsummer sunrise, set, Moon furthest north; view to east for different star?

14. Maryhill Road. Near Glasgow University Observatory. Good possibilities, between trees in some cases.

15. Dawsholm Park. Arcs to south and west, no north and east.

16. Garscadden Wood. High but lots of trees. From path could have distant settings, using flats etc. as foresights.

17. Lochgoin Avenue. No Midsummer sunrise. Good arc for rest in south. Ground not flat.

18. Inchfad Drive. No Midsummer sunrise – probably all rest. Sirius doubtful. East obscured.

Glenmuir Drive and Priesthill Road were the only places where no detailed assessment was attempted; in fact I told the driver not to stop. The visual impact of that area was worse than what I remember of the Gorbals before the great demolitions of the 60s, certainly far worse than, say, Easterhouse or Drumchapel, housing schemes which visiting commentators called "frontiers of hope-lessness" and the inhabitants called "deserts wi' windaes." To build a megalith in Nitshill with public money would have been such an insult that one would be inclined to help the vandals tear it down again. That glimpse of urban deprivation brought home the extent to which job creation and temporary employment schemes, whatever results they achieve and however valuable they are to the individuals employed, can only be viewed as window dressing at government level. The entire budget of the Astronomy Project would scarcely have renovated one of those buildings, and there were streets of them.

Even there the skylines might have had possibilities; I didn't stop to find out because it might have been difficult to give an adequate account of myself. At each of the other sites something could be done, although in some cases only a single alignment might be good enough to use. I also had to think about how the structure would look in relation to its surroundings, e.g., on Cathkin Braes and in Bellahouston Park. In Tollcross Park the football ground was functional rather than beautiful, and a stone circle would not be much of an ornament; the people directly across the road might not like having one planked right in front of them, presumably by heavy machinery. At Knowetap Street, which in other ways was quite good as its name implies, the road was so narrow that the structure would be almost in the gardens of the people opposite.

Finding Broomhill Park, as it was marked on my map, and finding a way in, gave the driver some trouble initially, for nei-ther he nor anyone he asked had ever heard of it. It was plainly marked on the map beside the M8 motorway, almost due north of the city center, but the only Broomhill anyone had heard of was out in the West End. It turned out that the park wasn't built yet and the site was still waste ground. After cruising the area the driver decided the only way in was up a steep track from Pink-ston Road; however this became impassable not far uphill, and I had to go on foot. Although this was the first site visited in the afternoon, I was still so wet that it hardly mattered. Taking the Miner's Dial seemed pointless, so I laid my master diagram over it in the car and took even rougher bearings on the Sighthill flats nearby before I set off.

Coming on to the top in the wind and rain, knee deep in long grass and rubble, was nevertheless dramatic enough to make the whole day. I was on the extreme east of a near flat hilltop, deep in trees and bushes, which was almost big enough to be called a plateau. The going was very rough where I was because of the foundations of earlier structures, but on the whole hilltop there was nothing standing except two or three square towers, two or three times the height of a man and built of wooden beams and corrugated iron sheeting, which proved to be the dovecotes of the local pigeon fanciers. What was absolutely amazing, since I was only three-quarters of a mile north of the City Chambers, was that I had an almost perfect true horizon right around the sky. The only major obstacles were the high flats, back over my right shoulder, and they could have spoiled one group of alignments at most. For all the rest of the surrounding 360° the skyline was almost entirely natural, yet low, and where buildings did rise above it they were still low enough for rising and settings over them to be acceptable. This was it! – if I could get it.

From the outset the omens were favorable. When I discussed the site with Paul Green, the Special Project Manager, and Frank O'Neill, the Shop Steward, photos were unearthed, and at first I had trouble recognizing them. Then we realized that the former Pinkston power station to the north had been demolished only months before, taking with it a huge cooling tower which previously had dominated the site and made it just an unimpressive piece of waste ground. Ken Naylor told me that planning the new park was the responsibility of Ian Clair, of James Cunning, Cunningham and Associates. When we met on June 2, he said that the highest point on the hilltop was being reserved as a viewpoint because of its dramatic view all over the city center; he was looking for an original idea and warmed immediately to the megalith proposal. For another propitious touch, he told me that to avoid confusion with the Broomhill in the West End, the Park would take its name from the ground to the north and the northwest, and be known as "Sighthill." Since megaliths, in turn, most often take their name from the locations, it could hardly be better.

Ian also started me on the research into the history of the site, giving me a photocopied map of the area "circa 1830" that eventually proved to be from the first edition of the 6-in. ordnance survey (1858). This enquiry was actually to lie fallow until after the project ended, because it went off down a side trail which proved eventually to be of considerable potential importance.

On the day before the helicopter operation, as the lunar stones were being put into position (see Chap. 7), up on the hilltop came a representative of the South of Scotland Electricity Board. He had come to see whether the Navy was still interested in landing on the former power station site to refuel. While we were discussing the next day's events, he looked around at the incomplete circle and said, "You know, this reminds me of the prehistoric one they took us to see in Drumchapel when I was a kid."

This riveted our attention because until then, we had no leads as to prehistoric circles in or around Glasgow. Our visitor could tell us only that the year was 1938 or '39, and the reason his school class was taken to see the site was because it was going to be built over with a petrol station. After the circle was built and the activity died down, I tried to pursue the lead with the late Chris Boyce, who was then a librarian with the *Daily Record*. Nothing came to light from the newspaper's files, and Chris suggested the Glasgow room at the Mitchell Library. I was by then too involved with the schools and exhibitions to take it further, but when Alan Maclean joined the project in autumn of 1979 I asked him to pursue the enquiry. Oddly (see below), Alan drew a blank at the Mitchell, whose staff referred him to the Ordnance Survey archives in Edinburgh.

Now a problem arose. The Astronomy Project had been cutting corners on established District Council procedures for more than a year in the drive to get results, and that was beginning to catch up with us. In particular there had been a lot of traveling outside the normal limit of the city boundary (see Chap. 6), unconventionally claimed back through petty cash, and now there had been a clampdown. If Alan had to go to Edinburgh he must put in for authorization and obtain a railway warrant. The warrant never arrived and the trip was never made. Later my first wife Linda went back to Glasgow Room at the Mitchell, on the off chance that something might have been missed.

The results were dramatic. What went wrong during Alan's visit remains a mystery, but there was in fact so much material on the prehistoric circle and on Sighthill that Linda could

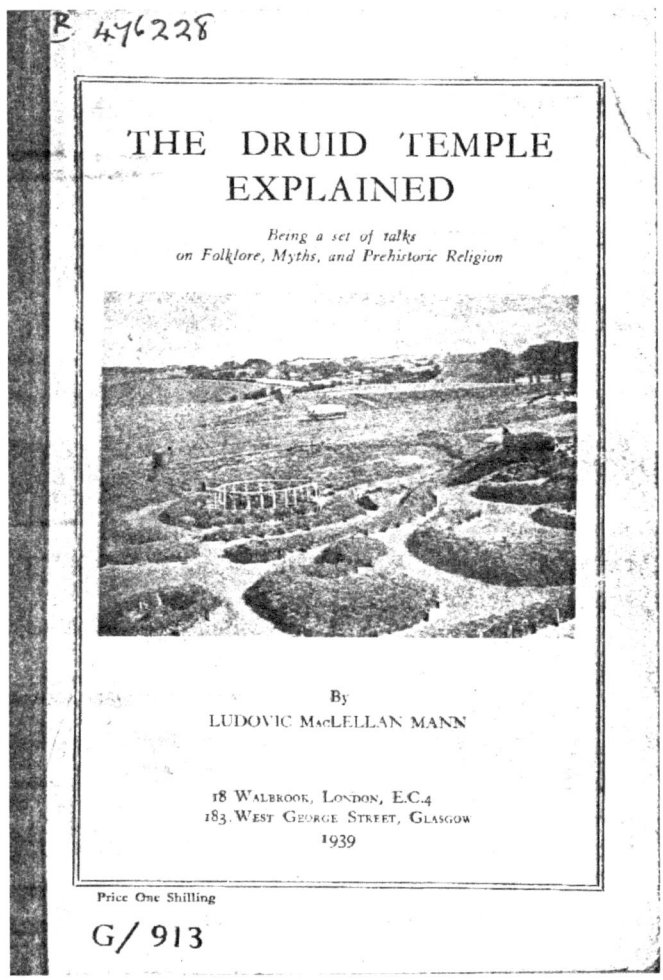

THE DRUID TEMPLE
EXPLAINED

*Being a set of talks
on Folklore, Myths, and Prehistoric Religion*

By
LUDOVIC MacLELLAN MANN

18 Walbrook, London, E.C.4
183 West George Street, Glasgow
1939

Price One Shilling

Fɪɢ. **5.7** Reconstruction of the Knappers structure at the 1938 San Francisco Exposition

take only preliminary notes for me (the Mitchell does not permit borrowing). The prehistoric circle was at a site called 'Knappers' in Clydebank, not Drumchapel, but with obvious flint-working connotations. It was a major tourist attraction in the 1930s and was visited by 40,000 people in 1938 [10].

A full-size replica of its central portion was erected at the San Francisco exposition (Fig. 5.7). It lay between the launching site of the *Queen Mary*, where the river Clyde bends to the north, and the Boulevard, the extension of Great Western Road that runs

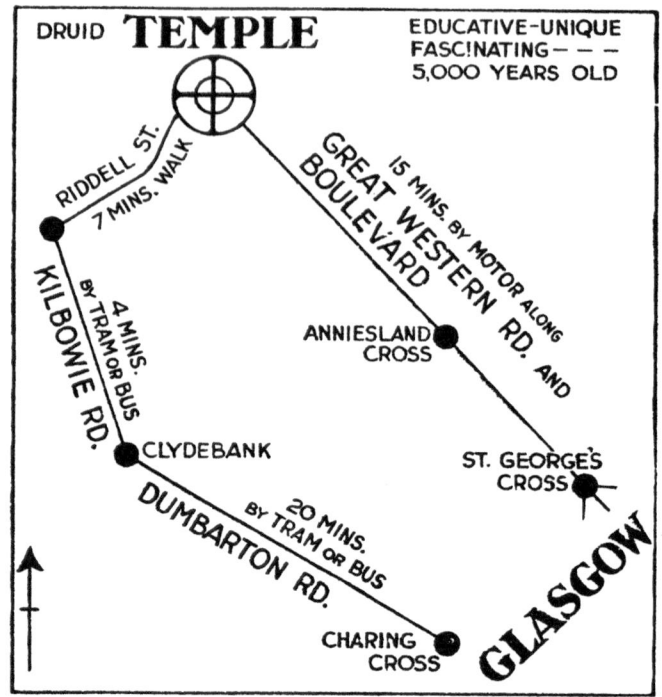

FIG. **5.8** Knappers location, back cover of "The Druid Temple Explained"

out of the city as the main road to Loch Lomond and the Highlands (Fig. 5.8). The sandy ground had formed as a shoal in the tidal estuary of the river, and the prehistoric remains were found in mid-July 1937 as sand was being excavated for building purposes. On the eastern side part of the structure was destroyed by the sand digging before its importance was recognized.

Archaeologists had excavated about a third of the total area by late summer of 1938, and work was resumed in summer of 1939; by then, however, "it is feared that unless sufficient funds are subscribed without delay that the temple may have to be abandoned by the Explorers, which would be an unforgivable national disgrace. Donations, however small, should be sent to Mr. J. Eric Ferguson, chartered accountant, 166 Buchanan Street, Glasgow" – down the road from our office. Evidently the appeal was not a success or time was too short, since it must have been in the same year that our SSEB informant went to see it before its destruction.

In view of the description that followed, it is amazing that the site should have been destroyed and even more incredible that the memory of it should have been dropped out of the archaeological literature. Part of the explanation is that Ludovic MacLennan Mann, the author of the pamphlets in the Mitchell Library, had a falling out with the other Glasgow archaeologists. The onset of the war may have been another factor. But it is still utterly astonishing that not a single reference work so far consulted mentions a site that was at least 300 ft across and seems to have included the major features of Stonehenge, Stenness, Avebury and the great wooden henges.

Mann went along with the prevailing mythology in associating the structure with the Druids – the title of the second leaflet was "The Druid Temple Explained" – and much of his effort was spent in finding fanciful connections between the layout of the site, adjacent cup-and-ring markings (of which there were a great many), local place names and Celtic mythology. As a result he was led into a quite extraordinary extension of the Celtic/Druid period, since he believed the circle dated to around 3000 B.C. His reasoning was based on his interpretation of the carving as depicting the solar eclipse of 2983 B.C., but having come by that route to a date soon after Stenness and the ring of Brodgar, he recognized that the alignments would have to be corrected for the change in the obliquity of the ecliptic (see below). The main part of the structure was built 800 years or so later, contemporary with Stonehenge II or Stonehenge III, but the Clydebank site, known as "Knappers" was so elaborate that its construction may well have taken as long as Stonehenge itself. Its period of active use seems to have gone on for a great deal longer because apparently it was unique among mainland sites in being used for systematic, individual burials.

At the center of the site there was a spiral stone structure, built around a rectangular chamber. Mann argued that because there were no burials within it the structure was an altar. (It might also have been for lying in state or cremation. The exit from the spiral faced midsummer sunset but was only a foot wide, so it's easier to imagine it as an elaborate altar rather than as a roofed building. If so, however, the Avebury Coves are crude by comparison.) Around it, with a diameter just twice that of the inside of the stone spiral, there was a wooden structure with a wide opening, equal in width to the radius, and with a "doorway" skewed about 7° to the west of south – not nearly enough, however, to point to moonset at further south. It reminded Mann of the Bluestone horseshoe at Stonehenge, although that's oriented to midsummer sunrise. "At its western side the horseshoe structure has been pressed inward by storm action, indicating that the upright posts were once lintelled" – in other words, like Stonehenge III, which numerous experts have said appears to have been built from a wooden model.

Around this real "mini Stonehenge" there was a double crescent of postholes, this time oriented slightly east of south – but again, not enough for its axis to be pointed to major standstill moonrise. It's tempting to suggest that the crescent and horseshoe were oriented to the rising and setting respectively of some bright southern star, but the entrances are of course much too wide for any precise identification to be made. But it's a red herring in any case because unless the observer stood on or in the "altar," it certainly wasn't for the midsummer sunset alignment. Although it's impossible to prove, it seems reasonable to suggest that the gaps are relatively so wide, compared with the Stonehenge Avenue, in order to allow the major standstill moonrise and moonset in the south to be seen from inside the linteled horseshoe beside the altar (Fig. 5.9).

Now we come to an interesting point of interpretation. Mann maintained that the double ring of posts, and the many other similar structures outside it, had enclosed earthen banks, which were either preserved or recreated in the reconstruction for the benefit of the visitors to the site. This is very different from MacKie's interpretation of the concentric patterns of postholes at the wooden henges, that they were holding up a conical roof. In support of his contention, Mann declared that late in 1938 a virtually identical layout of banks had been found on unploughed land at Formakin in Renfrewshire.

The Clydebank "temple" was not strictly a henge because it had a central feature, but that's a modern distinction, and there's no guarantee that the megalith builders took it for an actual rule. Dividing up the interior by earth walls into little "rooms" would seem to support MacKie's contention that Skara Brae is a stone version of Neolithic astronomers' communal dwellings, but there is one slight problem. The astronomical alignments so far suggested do require one to be able to see out, presumably over the outer banks! Intuitively one feels that they should have been

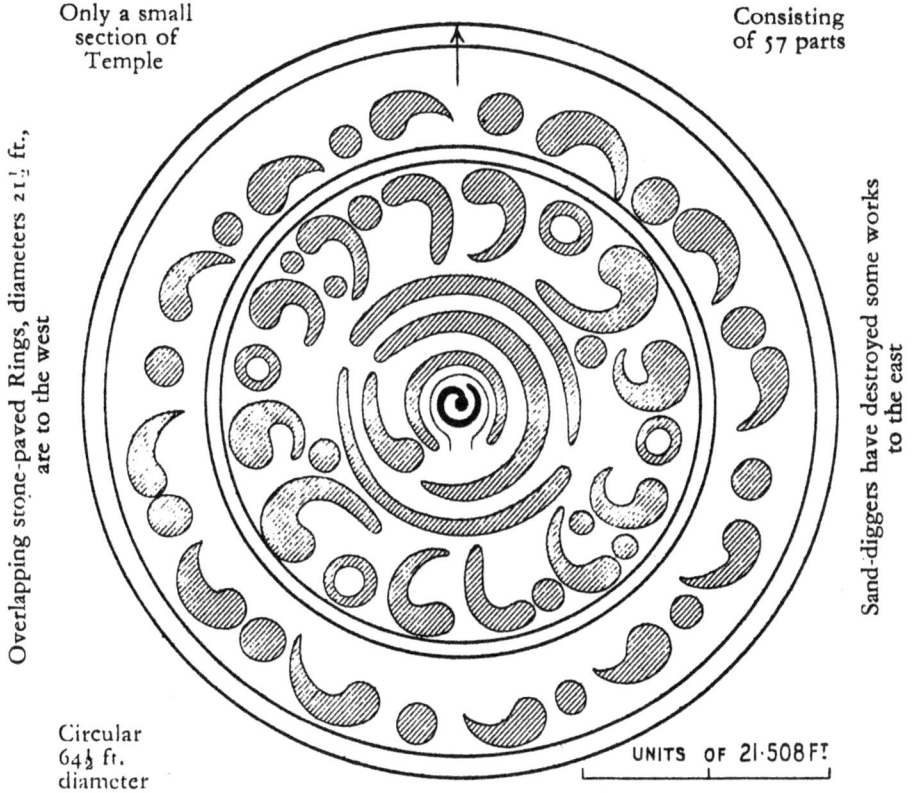

Only a small
section of
Temple

Consisting
of 57 parts

Overlapping stone-paved Rings, diameters 21½ ft., are to the west

Sand-diggers have destroyed some works to the east

Circular
64½ ft.
diameter

UNITS OF 21·508 FT.

Maze of low earthen mounds with marginal rows of wooden
stakes, near the centre of Druid Temple

FIG. 5.9 Knappers central plan, from "The Druid Temple Explained"

"activated" under an open sky, not within the gloom of a building
more than 120 ft across.

If we stick to the hypothesis of off-center viewpoints, however, we seem to be on the right track. If the inner sanctum was
solar and the inside of the horseshoe was for major lunar standstills, then it seems indisputable that the next ring out was for
minor standstill events (Fig. 5.9). There may even be support for
this interpretation in folklore. Earth mazes, known as "Nine Men's
Morris" or "Troy Town," were used in rural summer solstice rituals until Elizabethan times, at least [11].

Mann believed that the posts projected about 18 in. above
the ground. Their bases were firmly packed with clay and small
stones. At least 3,300 pieces of dressed timber were used on

the site, which yielded hundreds of stone implements and pottery fragments "both of the new Stone Age and the Bronze Age." Most of the tools were of white quartz, only a few of flint, despite the name of the site, which means 'flint-workers.' Hundreds of graves were found beneath the "flooring," by which Mann means artificially smooth prehistoric soil surface. From the variety of pottery types found, Mann estimated that the site had been used as a cemetery for no less than 2,500 years! Up to mid-1939 only a few graves had been opened, for fear of damaging the layout, so no one knows what happened to the rest. The graves clustered right up to the 'altar,' and Mann took that to be more evidence that the site was open to the sky: "Some curious features in the architecture of the Temple graves have not yet been fully explained. While three of the vertical walls of the shaft graves were sometimes built of stone, the fourth side was paneled in wood, which is now very much decayed. It does not seem that this was done to save stone material by substituting timber. Both materials were equally common and both were easily manipulated. Did wood more readily than stone afford access, out and in, of the migratory soul?" ("The Druid Temple explained", p. 11.)

There's no obvious answer to that question even now, but it's extremely interesting because in preparing a socket for a stone the megalith builders often braced three sides of it with thin branches, to keep the stone from crushing down the sides as it was tipped upright and dropped into the hole. Why the same principle should be used in graves in imitation is hard to see. Certainly it wasn't to take the shock of coffins!

The disposition of the skeletons showed that at Clydebank as elsewhere, the bodies had been exposed until disintegration set in, and some were buried in panniers. Others were cremated. The grave shafts were capped with stones, some of them with a central shaped block with a carved egg shaped cavity on top. Beads, tools, and round-based wooden and pottery utensils were found in the graves, and in one "a small, artificially shaped stone generally thought from its size and subdivisions to be (a) gauge for the old standard linear units used in all parts of the Temple structure and its layout. Perhaps the grave was that of the architect of the Temple. Gauges of the same character had been found in other ancient places in Scotland."

Have they so? Research has not revealed mention elsewhere of any such thing. Mann assumed his readers to be so familiar with "the standard linear units" that he didn't need to specify them. Notes on his diagrams implied that the banks were laid out in units of 21.508 ft, measuring from the center to the outside of the second ring bank, and from there to the outside of the next ring bank, and from there to the outside of the fourth. He also refers once to a "beta or small measure" of 18.9 ft; the difference, 2.608 ft, is close enough

Radii	Circumferences	
'Altar' inner ring	2.0 ft (0.74 MY)	0.28 ft (2.31 MY)
'Altar' outer ring	2.75 ft (1.01 MY)	8.64 ft (3.18 MY)
Horseshoe	4.25 ft (1.56 MY)	13.35 ft (4.91 MY)
Crescent inner ring	6.5 ft (2.39 MY)	20.42 ft (7.51 MY)
Crescent outer ring	9.25 ft (3.4 MY)	29.06 ft (10.68 MY)
(N.B. The posthole crescent as drawn by Mann is extremely rough if intended to be circular.)		
First ring bank, inner	13.5 ft (4.96 MY)	42.41 ft (15.59 MY)
First ring bank, outer	16.0 ft (5.88 MY)	50.27 ft (18.43 MY)
Second ring bank, inner	19.0 ft (6.99 MY)	59.69 ft (21.95 MY)
Second ring bank, outer	21.5 ft (7.9 MY)	67.54 ft (24.83 MY) (9.93 MR)
Third ring bank, inner	43.0 ft (15.81 MY)	135.09 ft (49.66 MY)
Third ring bank, outer	46.07 ft (16.94 MY)	144.73 ft (53.21 MY)
Fourth ring bank, inner	64.5 ft (23.71 MY)	202.6 3 ft (74.5 MY)
Fourth ring bank, outer	67.57 ft (24.54 MY)	212.23 ft (78.05 MY)

to the megalithic yard to be interesting. In fact, at 6.95 and 7.9 MY respectively, they are both highly interesting.

Let's now look at the Knappers dimensions overall in this context.

These values were taken from Mann's plans (Figs. 5.9 and 5.10), as published in his booklet. The measurements were made

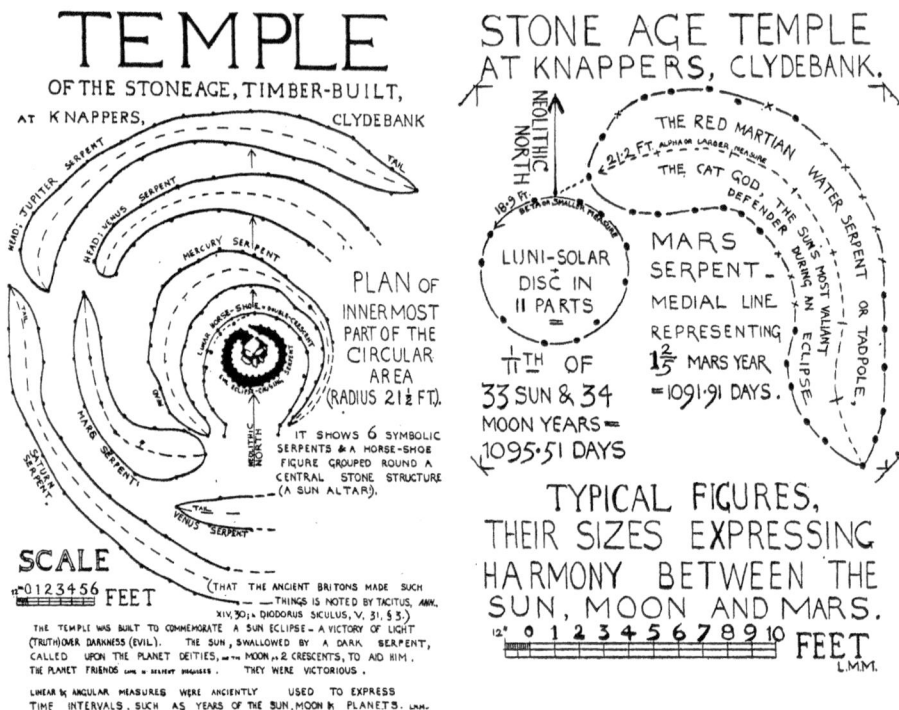

FIG. 5.10 Knappers dimensions from "The Druid Temple Explained"

straight across the plans from east to west, and treating all the features as circular, i.e., without rotating the ruler to find a diameter that would be most favorable to the megalithic yard hypothesis. It turns out that the "altar" was just over 2 MY across and the circumference of the concentric figures are close to the whole or half-multiples of the megalithic yard, as are many of their radii. It's interesting that the boundary between the inner structure and the outer rings should have a perimeter so close to 10 megalithic rods. And it's particularly interesting that these measurements are taken from plans published 18 years before Thom's *Megalithic Sites in Britain*.

A curious coincidence is worth mentioning in relation to Burl's claim, "It was a time when there were no months or weeks." On the Gate of the Sun at Tiahuanaco in Bolivia there is a complex carving that is generally agreed to be a calendar. Because it shows the year as divided into 12 months of equal length, it is generally taken to be lunar, and the symbols used for the days appear indeed to represent the phases of the Moon. The odd thing is, apparently only 24 days are shown in each lunar month, instead of the 30 of the Egyptians and the Maya. This problem has led to some wonderful "solutions," such as the Bellamy and Allen one in term of the cycles of fire and ice in Nazi cosmology. In *Man and the Stars* it was suggested, tongue in cheek, that it was a ready reckoner for extraterrestrial visitors who lived a 28-h day; the numerology, though fun, proved absolutely nothing and was left as an exercise for the reader [12]. But is it not interesting in this context that the two rings of figures surrounding the Knappers inner structure contain 24 figures each?

On the perimeter, exact positions unfortunately not marked, there were "traces of a ring of 19 large pillar stones (the same number as in the Stonehenge Bluestone horseshoe and the North approach to Callanish – D.L.) set apart at equal distances." Then we come to the appendix, "Results of Latest Diggings, July 1939":

> Much of the Temple area still awaits examination. Test diggings in the still intact and unexplored ground indicate the presence of many curious structures.

> A circular dwelling 21 ft in diameter and probably of wood and wattle walls, with a substantial wooden central standard 7 in. in diameter, has been revealed. On the floor were found prehistoric pottery fragments and many domestic stone tools, chiefly pounders.

> To the West, some 215 ft (79.04 MY – D.L.) from the main temple builders, have been excavated circular stone pavements each 43 ft in outer diameter (15.81 MY; circumference 24.803 MY). These overlap like circular links in the chain, and constitute a unique feature. From their centers, which form small pavements, radiate rows of stone and timber pillars which mark quite positively points on the distant hill horizons where the sun sets and rises at the solstices and equinoxes, thus registering the important stations in the sun's yearly journey.

> The old alignments differ slightly from the modern owing to the precessional changes (in the obliquity of the Ecliptic – D.L.), the amount of which shift being about 10¾ in. to the east at a radius of 50 ft from the observer's station (1°.03, nearly two solar diameters – D.L.)

> Traces of fires, probably ceremonial, occur along the Midsummer alignment.

The Ordnance Survey sheets betray field junction points and other features on these old sighting lines over distances of many hundreds of yards.

And then they built a petrol station over it!

The disappearance of the site from literature is in some ways even more extraordinary. Just one minor aspect of it is that in Mackie's *Science and Society in Prehistoric Britain*, he presented a map to show that the wooden henges, which he suggests to have been the dwellings of the astronomer priests, coincide with the discoveries of "Grooved Ware [13]." Glasgow is unique on the map in having Grooved Ware but no henges. Although round bottomed ware was found in the few grave shafts opened first at Knappers, Grooved Ware was found elsewhere on site, and the late discovery of a "circular dwelling" completes the correlation [14].

Mann's emphasis on the supposed Druid connection was probably a factor in the eclipse of the discoveries. In the account above we have screened out some very strange stuff – sticking with the eclipse metaphor one might say 'obscuring matter.' For example the feature here prosaically called 'first ring bank' he interprets partly as 'Venus serpent' and partly as 'The Red Martian Water Serpent or Tadpole' (Fig. 5.10). Who would not have paid to see *that* in front of the *Viking* Lander cameras?

Earliest Glasgow, another pamphlet of Mann's held by the Mitchell, is full of wonderful derivations of place-names from Celtic mythology. Some of these would be gratifying if true, such as Kow-Caldennis or Cowcaldanys for Cowcaddens, near Broomhill, and translated as "the grove of the Moon goddess." But a cautionary note is sounded when he identifies Rottenrow, the oldest road in Glasgow [15], with Routeneraw, hence Rath-na-era, "the earthwork of the Moon," rather than with the more modern Route de Roi, the King's Road, or Rattan Raw, the young women's road.

The Rath or earthwork in question, Mann suggests, may have been the Grummel Knowe, a prehistoric mound 'just a stone's throw from the Cathedral' and demolished in 1559 to surface roads. There was a second mound in Govan and a third near the Briggait called 'the Mutehill,' about which still less was known. As to the derivation of its name, he remarks only that it may be cognate with the adjectives Grim or Graham applied to the Antonine Wall. There is in fact more to be said. Many linear earthworks in England are known as Grim's Ditch or Grim's Dyke; the Neolithic flint mines in Norfolk, dating from 2500 B.C. or earlier, are known as Grimes Graves – with or without a mobile apostrophe. And there are many place-names such as Grimly, Grimley, Grimesthorpe. Two possible derivations are given by Brian Branston in *The Lost Gods of England* [16]. Apparently Grim, meaning 'having the face hidden by a hood,' was a nickname of the sky-god Odin and came from Old Norse 'Grimr.' But there was also the Old English *grima*, meaning a ghost, sprite or giant, and this is the derivation preferred by Ralph Whitlock in *In Search of Lost Gods, a Guide to British Folklore* [17]. In old Scots, we have the adjective 'grumly,' meaning ugly, with a definite connotation of evil, as in 'The Great Silkie of Sule Skerry' [18]:

> Then ane arose at her bed-foot, And a grumly guest I'm sure was he...

The *Glasgow Cathedral History* held in the Glasgow Room of the Mitchell unfortunately doesn't mention the Grummel Knowe. But it does stress that like many early churches the location was chosen as a deliberate challenge to the old religion. The Cathedral was built by the Molendinar Burn. Rituals allegedly were performed there and at a partly artificial amphitheater nearby. The statue of John Knox in the adjacent Necropolis, the City of the Dead, is claimed to be on a spot formerly used for Sun-worship. Apparently, however, the claim dates from the eighteenth century and has no historical evidence to back it up [1, 19A].

Mann doesn't speculate on the origins of another very interesting feature, Dobie's Loan, which ran in the first part of its course almost due west from the Cathedral site, then northwest past Broomhill. An enactment of the Magistrates and Council in 1655 declared that "the gerse of the Simerhill and lone that passes to Woodsyd, not to be selt heirafter, bot to lye for the use of the toune kye allanerlie, and non others to cleame right thairto." An article published in 1906, quoting "a *News* contributor," gives the route of the old drover's road as being along St. James's Road, passing the southern slope of the Summerhill as Dobbie's Loan; "the next turn taken by this historic loan is North Woodside Road, which winds up its course in University Avenue [19B]."

There are some odd things about that. Going southwest North Woodside Road dips right down to the river Kelvin, runs along it as South Woodside Road, then climbs to the west as Gibson Street. University Avenue itself climbs steeply, passing near the top of the Gilmorehill to which the university moved in 1864. It seems a very difficult route for cattle.

"Plan of the City of Glasgow, Gorbells, Caltoun and Environs, with an Exact Delineation of its Royalty, from an actual Survey by James Barry, 1782," confirms that St. James's Road is a modern name, and Dobbie's Loan, with its present spelling, was then the name all the way from what is now Castle Street, i.e., from within yards of the Cathedral. It also shows that the present southwest turn to meet Garscube Road is a detour forced by commercial development around the Glasgow Branch of the Forth and Clyde Canal. In 1782 the road went south around the bend of the canal, possibly following a contour, but then turned northwest again, crossed Garscube Road at a staggered junction and went on with the label 'Road to Kelvin Ford' (Fig. 5.11). In that section it may well have coincided with the appropriate part of North Woodside Road, and if the drover's road turned south again it would have entered Byres Road, logically enough. What is intriguing is that in prehistoric times, the line of Dobbie's Loan would have led from the Grummel Knowe towards midsummer sunset, to Broomhill and then straight for Knappers at Clydebank.

'The Simerhill' mentioned, also spelled 'Symmerhill' or 'Summerhill,' was also known as the 'Hill of Cowcaldanis,' and was a large hill, now largely obliterated by the canal and the motorway. 'The Hills of Glasgow,' a 1906 article, redrew a seventeenth century print from Slezer's *Theatrum Scotiae* to show the Summerhill's relation to Broomhill and the original Sighthill to the north (Fig. 5.12). The name is relatively modern but derives from a midsummer festival that was held on it from pre-Reformation times and whose antiquity was unknown. It began with the riding of the marches by the "Provost, baillies, and counsale, and deikines [deacons] on horseback," followed by "the haill [whole] inhabitants, fremen [freemen], and burgessis, with their armour, on futte." Thereafter a court was held on the hill at which "the town's twa herds," the birlawmen, the drummer and "minstrells" or pipers were appointed for the coming year, and other "use and wont" municipal business

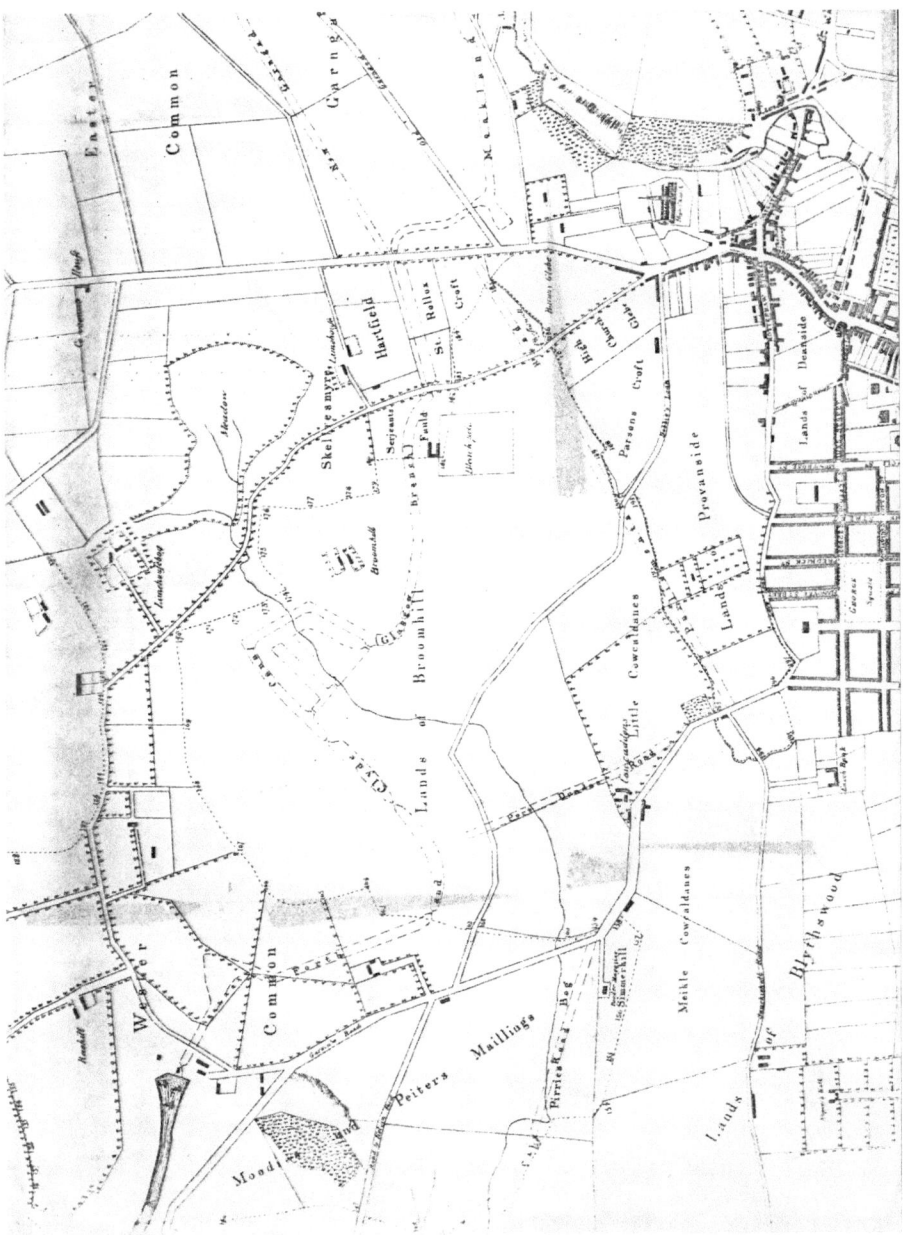

FIG. 5.11 Detail of 1782 map showing 'Lands of Broomhill.' George Square (present city center) at bottom *right*, Glasgow Cathedral ('High Church') at *lower right center*. Dobbie's Loan runs west from the Cathedral, then northwest to the Lands of Broomhill, and exits *upper left center* as 'Road to Kelvin Ford'

a

b

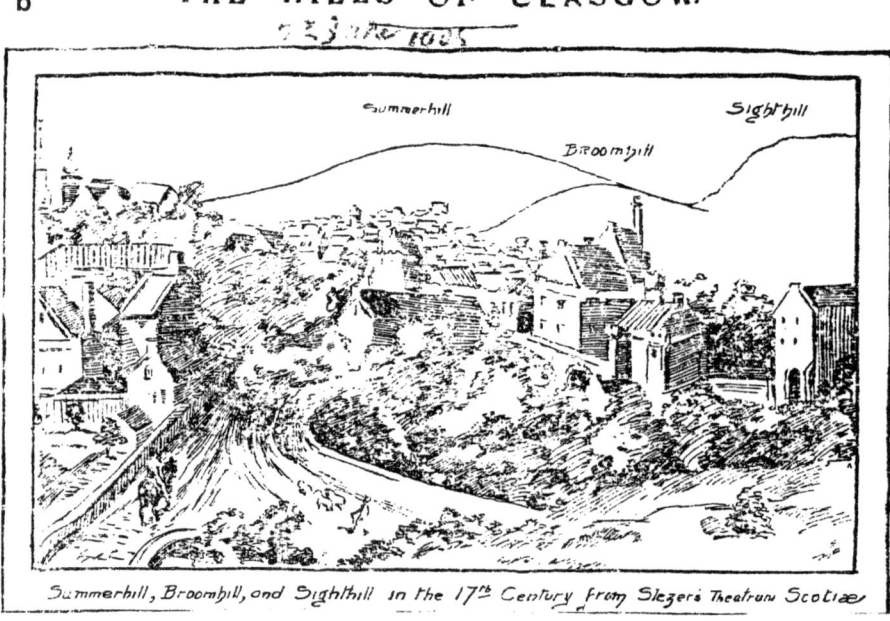

FIG. 5.12 (a) Detail from John Slezer, Theatrum Scotiae, 1693, looking north past Glasgow Cathedral on *right* [20]. (b) Summerhill, Broomhill and Sighthill marked on Slezer's drawing, 1906

was transacted. "In the afternoon cakes and ale, shinty and handba' filled up the time till evening, when milking the cows and putting the bairns to bed required attention [19B]."

The emphasis on the early halt to the proceedings is interesting, since in rural England midsummer celebrations notoriously continued into the hours of darkness. Since the original profile of the Summerhill has been destroyed it would be difficult to determine the exact viewpoint of the *Theatrum Scotiae* drawing, but it seems quite possible that the midsummer Sun would either have gone down the common slope of the two hills, or else set behind them and had a momentary reappearance at the foot of the notch. It would have been fascinating to know whether Summerhill marked the equinox sunrise, seen from Knappers, and whether its summit lined up with the ritual fires there. But it seems a reasonable guess that the revelers might originally have left the hill before sunset for another procession, down Dobbie's Loan to where the Cathedral now stands.

In recorded times up until 1590 the festival was held on Sundays, as if it had been 'Christianized' like Easter and midwinter solstice, but in 1577 there was added a 'wapynschawing,' the annual muster and inspection of men under arms in the district, "to be held on the day of the Symmerhill," and in 1590, mild though the program seems to have been with its early bedtime, it ceased to be held on Sundays "for the better observation of the Sabbath day." Could that have been because the Grummel Knowe had been gone for more than a generation and another kind of "observation" was no longer seen by the Church as a threat?

The subsequent history of the hills was a race between the agricultural and industrial revolutions. They remained common grazing until the general fencing of lands around Glasgow in 1770, then becoming part of the 'Lands of Broomhill' attached to Broomhill Farm, whose buildings were on the hilltop just north of where the circle is today, and whose southern boundary was Dobbie's Loan. The 1782 map shows this situation, with the name 'Simmerhill' now relegated to a small patch of ground containing a powder magazine south of New City Road. Its alternative name had moved bodily south to 'Meikle Cowcaldanes' and 'Little Cowcaldanes,' farmlands attached to Cowcaddens Farm (with the modern spelling) on Garscube Road, from which the modern Cowcaddens district, street and underground railway station have presumably taken their name.

Other things were happening meantime, however. Proposals for a canal joining the rivers Forth and Clyde had been in circulation since the 1760s, with a great deal of argument about the route, depth and scale. Work began at Grangemouth in 1768 and by

1775 had reached Stockingfield in Maryhill, Glasgow. There were insufficient funds to push on to the Clyde, but in 1777 a basin was opened at Hamiltonhill, near what is now Partick Thistle's football ground. The basin quickly became a busier port than the Broomie-law on the Clyde itself, and a focus for industrial development in the area.

In 1786 Robert Whitworth resumed work with a government loan of £50,000 and took the canal on to the Clyde at Bowling, building locks and an aqueduct at Maryhill, which became one of the city sights. At the same time a new terminal was built at Port Dundas, named after the company chairman, on the south side of Summerhill. It had an oval basin, a granary, and a link called 'The Cut of Junction' from its east end to the Monkland Canal, begun in 1770 by James Watt to carry coal into the city but not completed until 1793 and subsequently deepened in 1807 [21].

However, there is an oddity. The 1782 map shows the full length of the cut, identified as 'Forth and Clyde Canal (Glasgow Branch).' Since it is shown dotted perhaps it is the proposed route, or work in progress; if so, interestingly enough, the same applies to Parliamentary Road, then across country but later to become one of the city's most famous pub-crawls – "A' the pubs in Parly Road." (By the late 1970s it had been so carved up by housing developments, etc., that when John and I left the hilltop at 2.20 p.m. one day, after working through a very hot morning and lunch hour, we couldn't get a drink anywhere!)

It's still odder that the 1782 map shows the Basin and Tim-ber Basin between Broomhill and Summerhill just as they survive today. Yet the Mitchell Library has another map, apparently later since it shows 'Port Dundas Bason (sic)' and 'Port Dundas Town,' yet with no sign of the wharves. This 'Plan of the Lands of Hun-dred Acre Hill, Broomhill & C., Lying in the parish of Barony and Shire of Lanerk (sic), The Property of Mrs. Rae Crawfurd of Mil-ton,' was drawn up for the auction of three plots of ground, of which the third lot was 'Whin Park' in the bend of the canal where the wharves were later placed. Perhaps it was at that auction that the canal company or other developers acquired the ground.

At any rate, the plan shows how the canal put an end to the huge Broomhill Farm so soon after the Lands of Broomhill were enclosed. For the information of prospective buyers it helpfully shows the original contours of the south faces of Broomhill and Summerhill and makes it clear that the canal has gone well inside both of them. The Lands of Broomhill had now been split up into a large number of 'parks,' and those south of the canal were soon to lose the name altogether.

The former Summerhill now had a separate identity as the 'Hundred Acre Hill Farm,' as which, according to the 1906 article, it became "The Most Famous Dairy Place in Scotland." "Prior to 1866, as many as 18,000 cows have stood at one time in Harvey's Byres on Hundred Acre Hill and grazed over a wide expanse of country to the north and west. The site occupied by the majority of the Byres is now covered by Hugh Baird & Son's malt houses, but the asphalt floors of several byres are visible on the hill, immediately to the west of the malt-houses. Harvey's dairy farm still flourishes in a part of the old premises, and milk from various country districts is handled. The lowing of hundreds of cows is now an unknown sound in Hundred Acre Hill. When the cows left they took the name with them, and the rather foney (sic) appellation of Dundas Hill has taken its place [19B]."

There was one place, however, where the name lasted at least 28 years longer. In "The Bevvy, the Story of Glasgow and Drink," Rudolph Kenna records, "In 1894, the Hundred Acre Inn, Port Dundas Road – continuously licensed since 1799 – was still going strong. One report stated: 'The old kitchen with its venerable fireplace, and the parlour, where the canal men used to make merry, were still just as they were in the olden times.' Also in Port Dundas was the Old Basin Tavern – a favorite resort of canal boat passengers in the days of the 'hoolits' (owlets = night boats) playing between Glasgow and Edinburgh. In the 1890's, the pub was owned by the Great Canal Brewery. The Pear Grove Tavern, North Woodside Road, was a similar establishment – and also survived into the 1890s [22]."

The 1906 author goes on to argue strongly for the preservation of the hilltop, as a recreational area to accompany "a new experiment in workmen's dwellings" for "the poorest class of workers" – "four-storey tenements, containing houses of two rooms, and the new artistic one-room house, a model of which was on exhibition at the People's Palace" – in other words, the infamous 'but 'n' bens' and 'single-ends' that were to give the Gorbals the highest population density in Europe. "The outlines of the hill in many places are obliterated by the cuttings made for roads, streets, factories, and the canal, but the almost flat summit has never been built upon, nor materially changed in any way....Here is a suggestion that is perhaps worthy of serious attention from the Magistrates and Council. The open space known as Phoenix Square is situated on the slope of the old hill of Cowcaddens or Summerhill. Why not remove this park enclosure to the summit of the hill, where every westerly breeze blows freshly in from the country, and from which there is obtained A MAGNIFICENT VIEW of fields, hills and woods... [19B]"

Had this advice been taken, then our circle might have been still more appropriately placed in historical terms. But as Gavin's photographs clearly show, below, the skyline of what's left of Summerhill was instead crowned by a power station and by the White Horse Distillery, and only a short Phoenix Park Terrace commemorated the possibility. But see Fig. 10.12.

The original Sighthill was likewise lost to us. It seems quite possible that its name arose because its large, flat crest could have held a midsummer sunrise foresight with its flanking lunar standstill markers, for Summerhill. It has not been built on, exactly, but it is completely occupied by the huge Sighthill Cemetery or 'Northern Necropolis,' twinned with the one by Glasgow Cathedral. Ian Clair's plan showed an intriguing 'Martyr's Stone' north of Broomhill, but this proved to be modern and related to the Martyr Monument in the Cemetery – not to the religious Covenanters, though claimed by some of their supporters, but to the political radicals John Baird and Andrew Hardie, executed at Stirling in 1820 [23]. It did seem for a time that we might be able to use some of the more prominent tombs as foresights, but that was to lead to a great deal of trouble – see later.

Going back to the 1780's, however, the map shows another portent of big changes that were about to happen. On the 1782 one, there is a rectangular patch of ground on the lands of Broomhill, but extending on to the adjacent 'Serjeant's Fauld' and labeled 'Bleachfield.' It lay across what is now Pinkston Road from 'St. Rollox Croft,' through whose three fields the canal and Parliamentary Road were dotted in. Though the road hasn't yet come through 'Mr. Pinkston's property,' the Croft is now simply 'Mr. Glen's property'

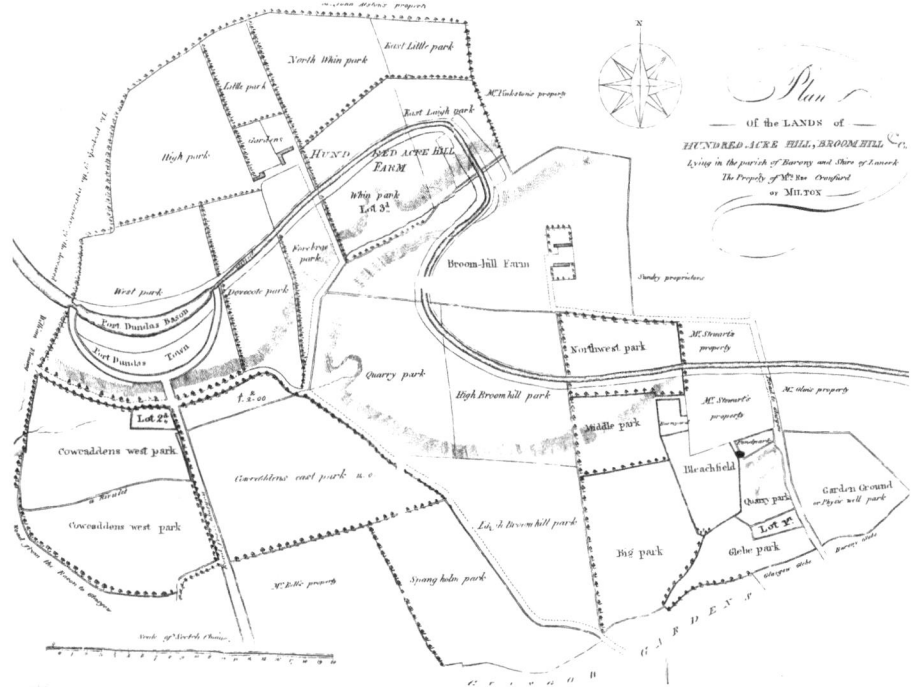

FIG. 5.13 Plan of the Hundred Acre Hill Farm, undated

(Fig. 5.13). The Bleachfield has been expelled from Serjeant's Fauld, which is now 'Mr. Stewart's property' (on both sides of the canal), but it has extended southward and become a permanent feature that was about to claim a lot of territory for the little-known St. Rollox, a.k.a. St. Roche.

In those days bleaching involved prolonged exposure to the Sun, and the first official bleachfield had been set up at Gray's Green in 1729, "probably catering for the embryonic linen industry in the city," says Hume. By the late eighteenth century there were quite a number of bleachfields around the perimeter of the city: Hume has found examples at Wellmeadowfield in Pollockshaws, Bellgreen, Bellshaugh, Kirklee in Kelvinside, Springfield in Dalmarnock, Kelvinhaugh and Hogganfield in Millerston and we can add Broomhill to that list. This phase of the early textile industry is immortalized in one of the most beautiful Glaswegian love songs, of which Pete Shepheard collected an abridged but unusually detailed version in Perthshire in 1965:

'Where are you goin', my bonnie lassie,
'An' what you do I would like to know.'
'Kind sir,' she answered, 'I am but a bleacher
On Cochrane's bleach field near Kelvinhaugh' [24]

This practice was to be brought to an end by the invention of bleaching powder in 1799, by Charles Tennant. Experiments in the use of chlorine had begun in Glasgow in 1787, at the suggestion of James Watt, and Tennant was the first to make it available in a convenient solid form by passing chlorine over lime, at his new chemical works on the northerly part of St. Rollox Croft. According to some sources the ground was bought from J. and R. Tennant, but in fact it was sold by J. Pattison to Tennant, Knox & Co., chemical manufacturers.

Development and diversification followed rapidly. In 1801 the company bought a 12-horsepower engine from Boulton & Watt for grinding manganese dioxide and chalk; they pioneered various developments in the manufacture of sulfuric acid, and by the 1820s had six competitors in the region of Port Dundas, where the prevailing winds blew the fumes away from the city. One of the by-products of the bleaching powder was sodium sulfate, used to make sulfur by the Leblanc process. The soda in turn was used in making soap, and a range of sulfates and sulfites was also produced. By the 1830s to 1840s the plant was claimed to be the largest chemical factory in the world [21].

Most of the buildings were single-story structures of brick and rubble, with pantiled roofs for ventilation, short-lived due to corrosion or changes in production methods. However the 'New' Office Block on the Cut of Junction, built around 1830, survived until closure and demolition in the 1960s. Lead chambers, Gay Lussac and Gower towers were prominent, but the site was dominated by its chimneys, and 'Tennant's Stalk,' built in 1841–1842, was the tallest in the city at 455½ ft. It was quite an attraction, and excursions were run to see it, with the result that the rival Joseph Townsend's Chemical Works in Port Dundas put up a 'Townsend's Stalk' 468 ft high, to overtop it. The 1906 author wrote, "On account of the summit not having been covered by high buildings, like the other elevated spots of Glasgow, the very existence of the hill is unknown to many citizens...but from its south-east slope, Townsend's Stalk towers over the city, so there need be no excuse for ignorance about the position of the ancient Summerhill [19B]." Townsend's Stalk was demolished in 1927, so there's more excuse now.

The name of St. Rollox, however, had caught on in a big way. There was the St. Rollox Cotton Mill, St. Rollox Engine Shed, Flint Glass Works, Ironworks (two of them), Railway Works, Spring Van and Lorry Works, Bottle and Flint Glass Works, Cooperage and the St. Rollox Cooperative Society, with its warehouse on Lister Street. Among Glasgow pubs at the end of the nineteenth century, "Another quaint survivor was The St. Rollox Tavern, Castle Street, where sheep's head broth was served in dining rooms richly hung with old paintings and engravings [22]."

Strangely enough nobody seems to have attached the name to a street, so after expanding over a great deal of territory it virtually disappeared again. In *The Industrial History of Glasgow*, Hume considers that the most lasting monument, not just to St. Rollox but to the entire Glasgow chemical industry, must be the great mounds of alkali waste from the plant that were later grassed and built over [21]. (A manager named J. MacTear pioneered the reclamation of sulfur from such waste in the 1870's.) Hume hasn't specified where those mounds are, however, which limits their value as a monument! A dip in Pinkston Road, north of the Broomhill, is still called 'Stinky Corner' or 'Stinky Ocean' with good reason. There was some concern that the waste might have been dumped on 'our' hilltop, in which case it might not be able to take the stones; but exploratory probes by the contractors' earth-movers found only good, dark clay.

In fact the plant's most lasting monument is what is *not* there, because the Summerhill wasn't the only area to be reshaped by the developers. The canal had already been carved into well within the original Sighthill base, and a private branch was cut leading into the St. Rollox works in 1836. But another form of transport was coming to change the Broomhill still more.

In 1849 the Caledonian Railway completed a second tunnel under the hill to their Buchanan Street terminal, with a ventilation shaft going down from the hilltop to the midpoint. Both tunnels began with cuttings that went deep into the north side of the Broomhill. Just when the south side of the hill was modified remains uncertain, but the Garnkirk & Glasgow Railway built a Glebe Street Terminal there in 1831, on the end of a line coming in from the east past the St. Rollox Railway Works. By 1858 those were simply the 'Caledonian Railway Company's Works,' but the Glebe Street station had become the 'St. Rollox Depot,' with a great spread of sidings along the Cut of Junction. To accommodate these the 1858 map shows that the south side of the hill had been cut right back, making it much steeper. Broomhill Farm and Broomhill House still survived on the hilltop, but the remaining land between the cuttings on the north and the sidings on the south was, literally, only a patch on the previous lands of Broomhill.

In the 1880s a railway swing bridge was built over the cut at Canal Street, linking the depot to the goods lines in Port Dundas. The first experimental electrical tramcars ran from Springburn to Mitchell Street in 1898, and as the horse-drawn system was being electrified over the next 3 years the experimental power plant was replaced by the full-scale Pinkston Power Station on the northwest of the Broomhill. It was put there because of the good road and rail access for coal supplies, and to use the canal for cooling water. The first large reciprocating engines were soon replaced by steam turbines, themselves modernized later. The 'handsome' twin chimneys were a landmark, but the cooling tower built in 1952–1954 was the largest in Europe at the time.

The wonders of the area were passing, however. The Monkland Canal was derelict by the 1920s, and where it passed through the chemical works, the Cut of Junction was enclosed in a pipe. Tennant's Stalk was demolished in 1922 and Townsend's in 1927. The main Forth & Clyde Canal continued to carry light traffic until it closed in 1962, after which the power station stretch became isolated cooling ponds. The swing bridge at Canal Street and the keeper's cottage were demolished in 1963; the chemical works, including the office block and the modern sulfuric acid plant, in 1964–1965; and Glebe Street Station in 1967.

In that same year Buchanan Street Station was closed and the tracks lifted. In the 1970s the M8 motorway pushed its way through from Charing Cross, the northern carriageway taking the line of the canal past Broomhill; and finally the power station and its huge cooling tower were demolished only months before I found the site in 1978. All that remained of the farming and industrial area were broken foundations here and there, especially on the east end of the park, and the railway sleepers still heaped where the canal sidings had been torn up. For over a decade the ground had been clear, used only by the pigeon-fanciers and at night by glue-sniffing children. When the contractors, Sportsworks Limited, began the landscaping of the hilltop, they found thousands of empty adhesive tubes just under the grass.

The nineteenth century slicing away of the south side of the hill had created a dramatic drop, with an almost unobstructed view over the city center and south side to the hills beyond, and it was this spot – the highest on the hilltop – which Ian Clair had reserved for a viewpoint. If it wasn't the best place for the circle, then there might be problems: I would need to find out in short order just how many alignments were actually clear at that point, and whether any other point was so much better that it might justify a change in plans for the new park at such a late stage. Since the area assigned for the view point was towards the western end of the hilltop, the Sighthill high flats were northeast from there and posed a threat to the midsummer sunrise group of alignments.

Working on my own was becoming a problem, but now I had a lucky break. John Braithwaite, an old friend with an impressive mastery of astronomical instruments, left Templeton's Carpets and asked if he could join the project. Since his skills were highly needed and there had been no other applicants, the MSC again agreed to waive the normal requirement that candidates over 24 years old should have been out of work for a year. Out-of-work astronomers as such are not easy to find, after all, and John joined me on May 9.

For day-to-day work on and off the hilltop, we felt that something more portable than the Miner's Dial was needed, and at that we had a bit of a shock. District Council procedures were such that to procure a pocket compass, we would have to fill out a request/requisition; this would be processed and go to the main Parks Department offices at Trongate. If approved, it would be passed to the Finance Office, who would put it out to tender. When tenders had been received and evaluated, in about 3 months an order would be placed.

John and I excused ourselves and repaired to Charles Frank Limited, where John's father had worked for many years and John himself had worked in holiday jobs, also full-time when he first left Strathclyde University. Five minutes of fast talking secured us a pocket compass on indefinite loan, and we left with an irreverence for official procedures that was to last for quite a time. A sore trial, no doubt, to the permanent staff who had to deal with us, but necessary if we were to have any chance of completing the circle in the originally allotted time, of which we had only 7 months left.

For his next miracle, John produced a portable clinometer, a tiny instrument that wobbled at the least breath of air when suspended, but would let us roughly assess the height of any obstruction. One of many questions still to be settled was whether, if a horizon event was completely obscured by a building, we should set our alignment to where the Sun or Moon first became visible, or to where it should be on the actual horizon. No doubt the determining factor would be just how many alignments were obscured; one or two lined up with the temporary irritant of buildings might be permissible.

Both instruments had rings for attaching them to walkers' belts. I was interested to see that the compass was identical to one my father retained from the war, except that the case was plastic instead of metal. The total weight of both was only a few ounces, but John supposed I would insist he carry both, gathering the office staff to witness his badges of servitude.

On the hilltop we waded around in the long grass, climbing over rubble in places, checking out a total of seven promising-looking spots. The ground seemed deserted, but suddenly we were accosted by an elderly man who came out from behind a hillock and required us to give an account of ourselves, while his terrier gave us a very suspicious eye. Hearing what was planned for the area, he declared that he and his dog would be just as well pleased if we took ourselves and our new park somewhere else. They liked the hilltop desolate as it was, ruins and all, so they could walk in peace. It didn't seem a very good omen for the popularity of our megalith in the area, and John was sufficiently impressed by this strange encounter to commemorate it in song, to the tune of *Strawberry Fair*, viz:

'As I was going to Sighthill Park,
Singing, singing, megaliths on Rollox,

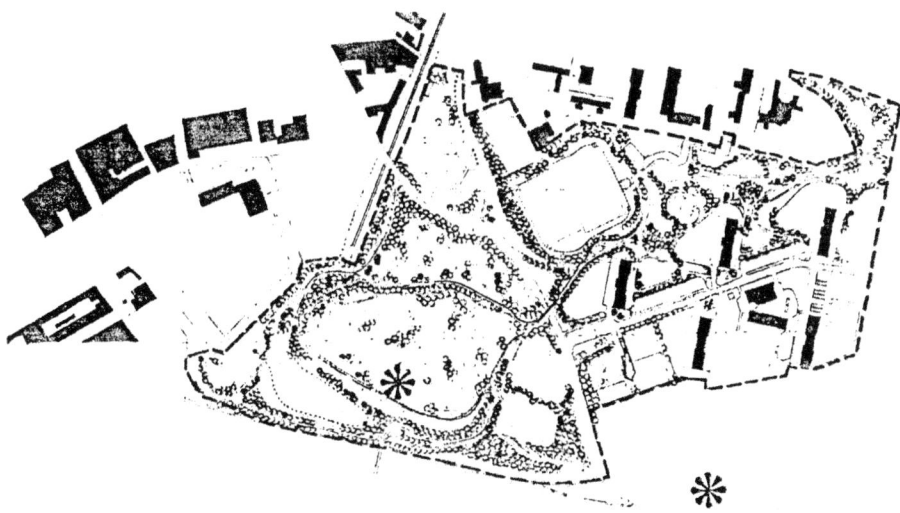

Fig. **5.14** Stone circle site in the new Sighthill Park, relative to the M8 motorway (at foot) and Springburn Road (at *right*). Pinkston Road runs through the Park Nor-Nor-West. Drawing by Dave McClymont, 1979 (Adapted from the Park plan)

> I spied a dog who was going 'Bark, Bark!' Fol-de-dee;
> And its owner said, "Just for a lark,
> You two have ruined our nice ruined park.
> Oh ri-fol, rl-fol, go and get 'im Fido,
> Ri-fol, ri-fol, fol di riddle dee."

That apart, it became clear that of the seven spots around the flattish top of the Broomhill, the highest surviving point, the one reserved for a view station, was indeed the best (Fig. 5.14). At most of the points midsummer sunrise was invisible behind the flats, but at Ian Clair's viewpoint it was well clear. It appeared that we would lose a lunar standstill alignment instead, but that was easier to live with, and we could set stones for it to emphasize the symmetry, arguing that the flats wouldn't be there forever.

Ian had also been worried about our proximity to the old Buchanan Street Tunnel airshaft. He didn't fancy having the weight of the stones near it or the possible hazard during construction. On the other hand, the shaft was to be permanently capped with concrete in due course, and perhaps the two operations could be combined. However, we found that the best place for the circle was well clear of the shaft and took maximum advantage of the view. Checking out several spots along the ridge, we settled on one where midsummer sunrise appeared to be marked by a pyramidal tomb in Sighthill Cemetery, capped with two square-shaped towers whose tips just met the natural skyline of the hills beyond. Since burial mounds were sometimes used as foresights it seemed appropriate, and as an indicator it was less ambiguous than the stone at Kintraw.

If we were going to work out accurate alignments in the office, from observations taken on the hilltop, we would need to be able to identify horizon features unambiguously. We needed a panorama of photographs that could be mounted on the wall in sequence so that the calculations could be related to it and no confusion would arise when we went back to the site. There was an official photographer attached to the Jobs Creation projects, but he had a very full schedule, so again we cut corners and appealed to Gavin Roberts, newly graduated from Glasgow School of Art, who had been co-illustrator of my first two books.

Having seen the site, Gavin wanted to start by photographing it from the outside, from Sighthill Cemetery, looking back along the midsummer sunrise alignment. At close quarters the tomb was extremely impressive, a cast-iron structure taller than we were, with inset panels of marble. Between its two towers was a great urn with the representation of a funeral cloth draped over it – this, too, all in cast iron. John pointed out the technological commitment this represented and how much it must have cost, as if nowadays one were to have such a piece cast in stainless steel.

Driving out of the cemetery, I realized that we had been so preoccupied with the functional purpose of our visit that we hadn't even noted the name on the tomb. "Walker," John declared confidently, and Gavin at once added that he must have been an armaments manufacturer to be able to commission a tomb like that. He and John immediately elaborated this into a right-wing fantasy in fake Yorkshire accents ("T' bonus? T' workers want t' bonus? I'll give them bloody bonus..") and we marched up on to Broomhill with considerable hilarity.

We hadn't been up there often enough yet to be absolutely sure of our spot, though it was at the top of a small gully that was to be the main approach, with the stones silhouetted against the sky – subject of a dramatic photograph by John Gilmour Fig. 7.27A. The gully had been strewn with empty beer cans since our last visit and immediately acquired the name of 'Dead Man's Gulch.' ("Do you think they'll hold fertility rites up here after we build a circle, Duncan?" "Of course, John, why should they stop now?") It was in this mood that we re-established our spot by taking a quick bearing on 'Walker's Tomb,' as we had all come to call it. After Gavin took the panorama, of course, the site should be fixed forever, since even a few feet of a difference would alter the relative bearings of near and distant buildings. Gavin therefore set up his tripod over a stone embedded in the ground on the line from 'Walker's Tomb,' and John ceremonially scratched a star on the stone to identify it for the future.

No sooner was this done, and Gavin barely started on taking the panoramic shots, than an extraordinary black cloud like something out of *Close Encounters of the Third Kind* came boiling over the western horizon and hurtled towards us as if aimed. No bigger than a man's hand when first seen, it was overhead in little more than a minute and sending down a ferocious downpour that covered the park and nothing else. Declaring 'I've started so I'll finish,' in the best *Mastermind* manner, Gavin refused to pack up, and we had to whip off our jackets to protect the camera – and also young Sarah, Gavin and Ruth's daughter, whose mother had not come out prepared for anything of this violence. The storm swept over us as fast as it had come, and Gavin did indeed finish – the menacing shadow could be seen in both the west and the east when the panorama was assembled (Fig. 5.15). As we retreated, soaked, it was suggested that 'Mr. Walker' be treated with more respect in future.

Finding out whether we did have the name right might be a first step, and on the next fine day when I was in the area, I went back to check. Walking into Sighthill Cemetery, even in bright sunlight, is a rather different experience from driving. It's a long way, with unrelieved perspectives of tombstones that weathering and pollution had turned mostly to a uniform and sinister black. The macabre atmosphere was enhanced when I rounded a particularly large tomb up on the summit and found myself among a jovial company of Hell's Angels having a picnic. Their bikes, all chrome and teeth, were parked among the tombstones, and the riders spread out on the graves with girlfriends, cans of lager and packets of chips, their studded jackets, emblazoned with skulls, swastikas, "Death" and "Heil Hitler," draped over the gravestones around. I passed among

Fig. 5.15 Viewpoint panorama by Gavin Roberts (1830 map to *left*), behind DL chair in the Astronomy Project office (Photo by Ian Downie, 1979)

them with an "afternoon" here and an 'aye-aye' there, took note of the information I had come for and left by another way.

The name on the tomb was not Walker (which was on another plaque two or three along) but Forrest. William Forrest (1839-1913) had evidently intended the great cast-iron structure to be a family vault, judging by the number of marble panels, but there were only three occupants. His son had died 24 years before him, at 4 years old, and his wife 3 years after him.

But with the joke over, nevertheless our problems didn't end. By this time we were exploring quarries in search of stones (Chap. 6), and on the day we went to Kilsyth, we had a flat tire as soon as we passed Sighthill Cemetery. The borrowed car had no spare tire, and we were stuck for the day. On the next attempt, as we passed the place where we were stuck, our naval friend David Proffitt remarked "we're all right this time" – no sooner said than another deluge fell on us.

Pressing on, however, we got to the stage where we would use a theodolite for a proper skyline survey. We began by consulting the Ordnance Survey map and picking landmarks that we could use for due north, but when we checked them on the panorama, John and I realized with a terrible sinking feeling that true north was a long way left of where we supposed. A hasty return to the site confirmed that we had made the classic mistake, subtracting the magnetic variation from the compass bearings instead of adding it. As a result, each of our rough alignments was 14° too far around in azimuth – nightmare visions of cutting cards to find out who would tell the Parks Department and who would tell Professor Thom! That at least had been avoided, but it would be bad enough to have to tell Clair at this late stage that we wanted to move the circle after all. John produced the 'badges of servitude' from his belt; but as we worked around the horizon, we realized with growing elation that now the alignments were better. All of them, it seemed, were in the clear, and most were on the natural skyline. We had been a great deal luckier than we deserved; and perhaps 'Mr. Walker' was after all trying to tell us something, and do us a much-needed favor.

> "The spirits then did make complain,
> Singing, singing, megaliths and morlocks,
> And Walker's ghost sent down the rain – fol-di-dee;
> It must have given him quite a turn
> When we said that the sun shone out of his urn.
> Oh, ri-fol, ri-fol, give the bums an eyeful.."

The time had come, however, to commit ourselves as to the basic design of the megalith. Obstructions on the skyline were no longer a problem. If one or two buildings were in the way we could ignore them, because obviously now we were going to mark alignments right around the clock. The Sighthill megalith would illustrate the full solar and lunar cycles. But how, exactly?

If we took our cue from midsummer sunrise at Stonehenge, the principal view point should be at the center with the markers on the periphery. The size and layout of Stonehenge III means that the view points for the lunar events had to be out by the bank, but that complication could be avoided on the relatively small scale to which Sighthill would be built. On the other hand, we had to consider the function of the circle as a monument in a modern park. If there was a central view point, perhaps with a plaque, then people would be pushing one another to look in the different directions, queuing up with their explanatory leaflets. An actual event such as midsummer sunrise, or still more a rare one like a lunar standstill, might find people actually fighting for the one and only privileged position!

But I already had another idea about Stonehenge that suggested an answer. The general interpretation was that given by Jacques Briard in *The Bronze Age in Barbarian Europe* [25]. "At that time there may have been great ceremonies, linked with a worship of the seasons and with agriculture, of which the high point was the awaited gleam of the golden day-star through the stones of the trilithons. The layout of the monument suggests a whole ritual: the arrival in procession along the avenue, the crossing of the sacred ditch and the separation of the faithful into different areas, only the initiated, the officiators at the ceremony or perhaps the princes of Wessex as well, being allowed to pass under the giant trilithons."

Briard had the beginnings of my idea in thinking about Stonehenge in terms of crowd control, but the layout didn't actually suit the ritual he described. The ditch and bank surrounding the circle go right up the Avenue, so it seems more likely that everything within them was forbidden territory to the uninitiated. If they were kept that far back, however, it would be very difficult to conduct a ceremony at the center. And even if they were allowed within the bank to surround the circle, relatively few would be able to see past the sarsens, bluestones and trilithons. Only the very few right at the center would actually see the *event*, the sunrise, in its intended relation to the structure, and that just doesn't make political sense in relation to the Apollo-type commitment of resources to building Stonehenge.

NASA had no intention of burdening the *Apollo 11* astronauts with a TV camera until Congress insisted on it – only one man could be first on the Moon, but the hundreds of thousands who worked for it, the millions who paid for it and the thousands of millions whom their leaders wanted to impress must all be able to see what happened. Since 'the princes of Wessex' had no TV remotes their ceremony would have been a spectacle, out where everybody could see it.

More recently, it's been argued anew that the focus at Stonehenge wasn't midsummer, but midwinter – see Chap. 4. Aubrey Burl pointed out in 1999 that the tallest stones of Stonehenge III are on the winter solstice sunset arc [26], and in 2003 it was suggested that the crowd stayed on the Avenue to watch midwinter sunset, and moonset at furthest south (every 18.61 years, for

12 months) through the 'windows' of the main trilithon and the sarsen arches [27].

Still more recently, a banked processional avenue 10 m wide, 'the first Neolithic road,' has been found leading to the river Avon from the nearby wooden structure of Durrington Walls, where there's evidence of vast feasts at midsummer and midwinter. The structure is oriented to midsummer sunset and midwinter sunrise, the opposite of Stonehenge. Where the Stonehenge Avenue meets the river a 'bluestonehenge' has been discovered, dated to 3000 B.C., so the thought is that people celebrated midwinter at Durrington Walls, then came along the river and up the Stonehenge Avenue at sunset, to inter their cremated dead nearby or even in the Aubrey Holes [28] (the pits named after John Aubrey, the Elizabethan diarist who first noticed them, and which Sir Fred Hoyle thought were an eclipse predictor).

Human cremations were found in the Aubrey Holes in the 1920s, but the excavators scooped them together and dumped them unexamined in Hole 7. They have now been exhumed for study. There were 60 burials in all [29], continuing until 2500 B.C. at least, and were almost all men, with just one woman among them. They were healthy apart from arthritis and may have belonged to a royal dynasty [30]. But Durrington Walls was occupied for less than 50 years, probably less than 35 years, which supports the alternative idea that burial in Stonehenge may have been an honor reserved for its builders [28].

It's possible that midwinter events took place as well, but since they suggest no function for the Heelstone, I'm holding to my vision of events at midsummer. It's interesting, then, that the ditch and bank go up the Avenue as if the watchers had to keep off it (Fig. 5.16). When that work was done the Heelstone was already in place, and it was given its own ditch and bank – so they had to keep back from that, too. That brings us to why the Heelstone is displaced from the center line (Fig. 2.4). The answer, perhaps, is that stones don't officiate at ceremonies – people do. On the spot at Stonehenge, I found it easy to imagine that the watchers lined the Avenue, outside the bank, while the leader faced the rising Sun beside the Heelstone, and his shadow fell back through the trilithon into the mysterious interior of the circle.

FIG. 5.16 Aerial view of Stonehenge, showing the ditch and bank extended past the Heelstone (Photo by Chris Stanley)

That's a subjective interpretation with a vengeance. Almost certainly it could never be proved, but that doesn't matter. It can't very easily be disproved, either, and it gave me the answer to the design question. Since we weren't planning to conduct any ceremonies, I could make a stone the center of the action on the sight-line; and since people would be able to see over it, on the scale I intended, I could make it a central stone. The view points could be on the perimeter, and people walking around the outside of the circle with the leaflets would be able to look over it from all angles without bumping into one another. What we would have, in effect, would be the ground plan of the view stations around Le Grand Menhir Brisé (Fig. 4.17b) – scaled down to 40 ft across and corrected for the latitude and date of modern Glasgow.

I showed the plan to Archie Roy, who approved it. Among the experts I had sounded out on the design question was E.C. Krupp, at the Griffith Observatory in Los Angeles, at Prof. Thom's suggestion (see below). At the point when my conclusions had been reached, I was most heartened to receive a letter from him that included the comment, "Of course, you could turn the whole thing inside out, à la Carnac [31]." As that was just what I had done, the support was extremely welcome and undoubtedly helped to convince the Parks Department that I was on the right track.

Once I had convinced the Parks Department and the Manpower Services Commission that we had to find a suitable site, then design a twentieth-century structure according to the ancient principles, I had turned my attack on the suggestion that modern materials such as steel tubing should be used. In fact Glasgow already had one such structure, the Steelhenge on the Strathclyde University campus. I had looked at it, and at photographs of structures in modern materials built by Chris Jennings, who had made extensive studies of standing stones (illustrating MacKie's books among others), but in both cases I felt the dominant allusions were to modern sculpture rather than ancient astronomy. To capture the spirit of the ancient sites, and produce a monument that would last for thousands of years, I felt that only stone would suffice.

I turned for support to Prof. Thom, whom I had never met. Ian Downie, who later joined the project, gave me his contact details, and I made an appointment to see him on April 13, at the house in Dunlop which he had built for his wife and himself during the Depression. I was warned that he didn't suffer fools gladly, so I knocked at his door with some diffidence.

> "Good morning," he said as he opened the door. "Have you seen the latest criticism of my work in *Antiquity*?"
>
> Not wishing to admit that I wasn't a regular reader of *Antiquity*, I said, "No, Professor, I haven't seen it yet."
>
> "Well, don't bother," he said, "the man's totally innumerate. Come in!"

Having heard what the project was, Thom was opposed to it. He had no wish for any kind of a tribute, and in his view the money would be far better spent on preserving ancient sites. While I fully saw his point, I had to point out that the money was allocated to build something of an archeoastronomical nature, and it had to be within the Glasgow city boundary – and all the ancient sites in that area had been destroyed. I did however get his grudging agreement that if we were going to do it, stone would be far better for the purpose.

Fig. 5.17 Viewpoint from the M8 footbridge, looking north (Photo by Frank O'Neill, May 1978)

At this stage Frank O'Neill, the NALGO shop steward, had to make his quarterly report to the Director of Parks on the progress of the various Special Projects. John and I took him up to the site and to show the location Frank photographed it (and us) as we went over the M8, illustrating that the footbridge led straight towards it (Fig. 5.17). On the hilltop, I had Frank photograph John in the position of the central stone, while I took the part first of the midsummer sunrise marker, then midwinter sunset. Combining these by the wonders of modern science (Fig. 5.18), we had the basis for an artist's impression.

Gavin Roberts was being increasingly drawn into the project, and we now asked the MSC if he could be appointed as our assistant. He had not been out of college and out of work for the statutory period (6 months in his case), but once again an exception was made since there were no other applicants. Meantime Bill Jones, a draughtsman working under Paul Green and sharing Paul's increasingly crowded office with us, had produced a preliminary plan of the circle using the perfect theoretical alignments. This was to use as a reference in the survey and also to show that we were well clear of the railway tunnel airshaft. Using that and

Fig. **5.18** Midsummer and midwinter alignments. Royal Infirmary and Glasgow Cathedral spire at rear (Photo by Frank O'Neill)

the composite photograph, Gavin beheaded John and me, replaced our trunks with stones, and filled in the rest to generate an artist's impression of the circle seen from the west. Bill's plan and Gavin's impression would later be used in our application to the Planning Department for approval (Chap. 7).

Copies of these were forwarded at once to Ian Clair, who approved them and asked if we could supply a model to be incorporated into the Consultants' official model of the park. It was not a tall order, anything but, since it would have to be about ⅜ of an inch across. In fact Gavin had to exaggerate the scale slightly, but he succeeded in building it in India-rubber, and before handing it over he photographed it from above with an anglepoise lamp simulating midsummer sunrise (Fig. 5.19). It was very unlikely that we would get such strong shadows in modern times – unlikely that they were so clear even in the Neolithic – but the model very clearly illustrated the principle of the layout. Standing on the side furthest from the Sun, the observer would position himself so that the central stone obscured his view of the midsummer sunrise marker, and for him, the Sun would rise over the central stone. Others standing around would see the shadow of the marker fall on the central stone and the shadow of the central stone fall towards the viewpoint. That little model photographed on Gavin's carpet was to prove enormously helpful, and months after the circle was

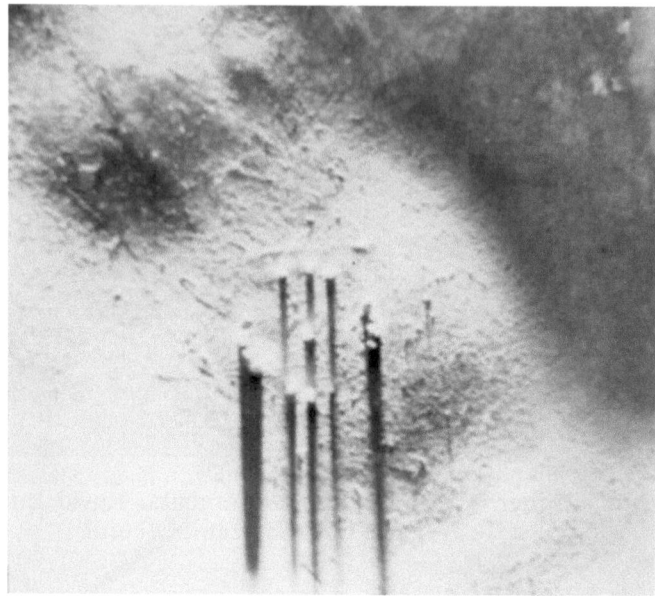

Fig. 5.19 Megalith model showing midsummer sunrise alignment (Model and photo by Gavin Roberts)

built, my then brother-in-law looking at the photograph was to remark, "You know, I've never understood what you meant by an alignment until now."

References

1. Saunders, D.: The Glasgow Diary, by Donald Saunders, Mostly. Polygon Books, Edinburgh (1984)
2. Thom, A.: Megalithic Sites in Britain, op cit
3. Krupp, E.C.: Echoes of the Ancient Skies, the Astronomy of Lost Civilizations, op cit.
4. Atkinson, R.J.C.: Stonehenge and Avebury, op cit; Crompton, P.: Stonehenge of the Kings. John Baker, London (1967); Harry Harrison and Leon Stover: Stonehenge. Peter Davies, London (1972)
5A. Roy, A.: Saturday Extra: Prehistoric lifestyle in doubt. Glasgow Herald, Sep 27, 1975
5B. Evans, A.C.: The three dimensional grid. Paper presented at 'Heresies in Archaeoastronomy. Edinburgh International Science

Festival, 16 April 1995; Lunan, D.: Epsilon Boötis Revisited. Analog CXVIII, **3**, 52–68 (March 1998)

6. Clube, V., Napier, B.: The Cosmic Serpent, op cit
7. Steel, D.: Rogue Asteroids and Doomsday Comets. Wiley, New York (1995)
8. MacKie, E.: The Megalith Builders, op cit; MacKie, E.W.: Implications for Archaeology, op cit
9. Aldiss, B.: Guest editorial. Fantasy Sci. Fiction **54**, 4 (1978)
10. Mann, L.: The Druid Temple Explained. William Rudge, Glasgow and Edinburgh (1939)
11. History Zone: Mazes and Labyrinths. BBC-2 Channel, UK, 19 June 1999; Foot, S.: Happy to lose himself in those long and winding mysteries. Daily Telegraph, 27 Aug 1994; Meaden, G.T.: The Goddess of the Stones. Souvenir Press, London (1991)
12. Lunan, D.: Man and the Stars. Souvenir, London (1974)
13. MacKie, E.W.: Science and Society in Prehistoric Britain, op cit
14. Ritchie, G.: Royal Commission for the Ancient and Historic Monuments of Scotland: Lecture, 'Knappers, Dumbartonshire, A Reassessment'. Archaeological Society of Glasgow, 17 Dec 1981
15. Mann, L.M.: Earliest Glasgow. William Hodge and Co., Glasgow and Edinburgh (1938); Renwick, R., Lindsay, Sir John: History of Glasgow. Maclehose, Jackson and Co., Glasgow (1921)
16. Branston, B.: The Lost Gods of England. Thames and Hudson, London (1957)
17. Whitlock, R.: In Search of Lost Gods, a Guide to British Folklore. Phaidon, Oxford (1979)
18. Benwell, G., Waugh, A.: Sea Enchantress, The Tale of the Mermaid and her Kin. Hutchinson, London (1961). Another version of the ballad is in Buchan, N. and Hall, P. (eds.): The Scottish Folksinger. Collins, Glasgow (1973)
19A. Glasgow Cathedral History. Glasgow Room, Mitchell Library; Holder, G.: The Guide to Mysterious Glasgow. The History Press, Stroud, Gloucestershire (2009)
19B. Anonymous: The hills of Glasgow. Glasgow Evening News, 7 June 1906
20. Kenna, R.: Glasgow Then and Now. Fort Publishing, Ayr (2001)
21. Hume, J.R.: The Industrial Archaeology of Glasgow. Blackie, Glasgow (1974)
22. Kenna, R., Sutherland, I.: The Bevvy: The Story of Glasgow and Drink. Clutha Books, Glasgow (2000)
23. Rogers, C.: Monuments in Scotland, vol. 1. Charles Griffin, London (1871)

24. Shepheard, P.: Collector's Piece: The Bleacher Lassie o' Kelvin-haugh. Chapbook, **5**(2), 10–11 (Christmas 1968)
25. Briard, J.: The Bronze Age in Barbarian Europe. Routledge and Kegan Paul, London (1979)
26. Stonehenge: Secret of the Stones. Channel 4, UK, 20 June 1999
27. Stonehenge: The True Story. Channel 5, UK, 5 Aug 2003
28. Secrets of Stonehenge. Yesterday Channel, UK, op cit
29. Oliver, N.: History of Ancient Britain Special: Orkney's Stone Temple, op cit
30. Time Team Special. The Secrets of Stonehenge, Channel 4, UK, 1 June 2009
31. Krupp, E.C.: Personal communication, 22 May 1978

6. Selecting the Stones

If ever you find yourself in a strange town and there's a quarrymasters' convention going on, try to gatecrash it: they're such great people.

– John Braithwaite

The next task was to find suitable stones, not so easy to come by in the twentieth century. We began by consulting Mr. McKenzie of the Department of Architecture and Related Services, who started us on a prolonged odyssey around quarries in the west of Scotland, with telephone inquiries going much further afield. Though the quarrymasters could not have been more cooperative, the problem was that almost all modern quarries used gelignite, which brings down the rock in very small pieces suitable for motorway construction and the like. Gavin's artist's impression had confirmed my view that we wanted stones to stand 4–5 ft 6 in. above ground – comparable, say, with some of the ancient stones on the island of Colonsay (Fig. 4.23). To bring down rock in pieces of that size would require slower burning explosive, probably black powder. We had the off-duty advice of the late LAEM David Proffitt of HMS Gannet, the anti-submarine and air-sea rescue helicopter squadron at Prestwick Airport. David was trained in black powder blasting and would have been prepared (given official permission) to do the job for us, but we still had to find a suitable quarry.

As John Braithwaite said, the quarrymasters were wonderful. One of our first visits was to a quarry near Dumbarton, and after we left, John said, "Do you notice how everyone is captivated by it? When we walked in he was prepared to dismiss us as a pair of nut cases, and now though he can't help us, he's going to find someone who can." At another, we were given the run of the quarry on the off chance that we might find stones big enough, "But we'll be blasting at noon, wherever you are." At a limestone quarry near Dunlop, not far from the Thoms, they would happily provide large

D. Lunan, *The Stones and the Stars: Building Scotland's Newest Megalith*, Astronomers' Universe, DOI 10.1007/978-1-4614-5354-3_6, © Springer Science+Business Media New York 2013

FIG. 6.1 Beltmoss (Back of the Hill) Quarry (Photo by author)

enough stones, but at great cost because they supplied a specialty market and would have to be paid *not* to shape them, to another order, once they had been extracted. They also gave us the free advice that limestone was unlikely to last, exposed on a hilltop (and indeed a later, symbolic limestone circle outside the Spaceguard Observatory in Powys, Wales, is already showing considerable erosion – Fig. 7.30). At Hillhouse Quarry, above my home town of Troon, we were given shelter during blasting in the quarrymaster's house, and when we met him, he was even prepared in principle to give us access to a virgin face and allow David Proffitt to blast it, though they would then have to diversify and find an outlet for black powder products, because it would be unsafe to use another explosive on it forever after. That one would potentially be so costly for the quarry, that we promised not to come back unless there was no other choice.

Finally, however, the search ended at Beltmoss quarry (Fig. 6.1), on Tak-ma-doon Road, Kilsyth (known locally as the Back of the Hill Quarry), a family business making curbstones and paving stones (Fig. 6.2), the last black powder quarry in Scotland if not in Britain. Kilsyth was originally Kelvesith, 'Kelvin-lands,'

FIG. 6.2 (a) The quarry at work. From here to the end of Chap. 7, photos by Gavin Roberts unless otherwise credited. (b) John asking questions. "I can see how you get the straight edges, but not how you do the curved ones?" "That's because we're masons, sonny"

FIG. 6.3 John with Mr. Motherwell and what was to become the central stone

i.e., those around the headwaters of the river Kelvin, which flows into the Clyde at Partick in Glasgow. The first commercial potato crop in Scotland was grown at Neilston Croft, on Tak-ma-doon Road, and there was a well dedicated to St. Mirren next to it [1]. Nowadays, the road runs across the golf course.

The quarrymaster, Mr. Motherwell (Fig. 6.3) and his foreman Jimmy (Fig. 6.4) gave us the free run of the quarry, and we picked out 18 whinstone blocks in the right size range, to which they spontaneously added five smaller ones as spares in case they should be needed. It wasn't as easy as that sounds. After drilling, placing and exploding the charge, very often the rock splits along a shear line, which leaves an artificial surface unsuitable for our purpose (Fig. 6.5).

FIG. 6.4 John and DL with Jimmy (Foreman)

FIG. 6.5 Stones with blasting marks and artificial faces

FIG. 6.6 Another unsuitable stone (John with paintpot)

To identify our stones, Gavin Roberts provided a paintbrush and a pot of black paint (Fig. 6.6). He was anxious that the crosses shouldn't wear off before the stones were extracted, unnecessarily as it turned out, as blackboard paint is virtually indestructible. After erection at Sighthill, the black crosses on the stones were to remain visible for years.

The first stone in the sequence was ceremonially claimed by my first wife, and as project manager, I ceremonially claimed the second (Fig. 6.7a). John claimed the third after some question over whether it was the right one (Fig. 6.7b, c). But after that, as access to the stones we wanted grew more difficult, I was happy to defer to John's expertise in rock climbing (Figs. 6.8, 6.9, 6.10, and 6.11), while Gavin continued to document the process, alternating between black and white and color.

Gavin had decided he would follow the bowed stone of Fig. 6.11 through the entire process from selection to completion of the circle. The circle is not completed to this day, but the career of the bowed stone was to be much shorter (see below).

Figure 6.11 also illustrates the biggest difficulty we encountered as we worked our way around the quarry. Having been brought down by blasting, many of the slopes now consisted of

FIG. 6.7 (a) DL claims stone. (b) John in doubt (Neolithic Man surfaces).
(c) John claims stone

FIG. 6.8 Over to John

FIG. 6.9 John at height

FIG. 6.10 John at greater height

FIG. 6.11 The bowed stone

Fɪɢ. **6.12** (**a, b**) "How are we going to get up there?"

highly unstable scree, far from its natural angle of repose. The problem came to a head with two attractive paired stones that we really wanted, but John could see no way to reach them single-handed (Figs. 6.12 and 6.13). Despite my lack of experience there was no option but for me to go up with the paint, while John took the brush (Fig. 6.14). At the crucial moment I held out the pot and John swung around, dipped the brush and painted the fastest cross in the history of art, as his feet slid out from under him (Fig. 6.15).

Our last target, the future central stone, was unreachable by any means. Among his many other skills, however, John had been Junior Javelin Champion for Scotland, and casting about, surprisingly we found a stick into which the brush could be wedged to improvise one (Fig. 6.16). And John hit the stone, first time (Fig. 6.17), but the brush left no mark, so there was no option but to designate the stone with an arrow (Fig. 6.18).

Fig. 6.13 John tries the direct approach

Fig. 6.14 The joint attack

Fɪɢ. 6.15 The fast cross

Fɪɢ. 6.16 The javelin

FIG. **6.17** First strike

Gavin's misgivings about the crosses wearing off were doubly misplaced. As soon as we had marked up the stones, Mr. Motherwell and his staff brought in a JCB with an articulated bucket, and in an amazingly short time they had all the stones down and gathered together (Fig. 6.19). Behind our stones, as they were waiting for uplift and transport to Glasgow, what would have been their fate otherwise could be seen – the flagstones that were the quarry's regular output (Fig. 6.20a). The stones were picked up by bulldozer and loaded into dump trucks, including the bowed stone, and Gavin traveled with them into Glasgow, where they were deposited at Ruchill Park for safekeeping. Unfortunately, their weight was such that the dumpers had to be tipped right up to dislodge them, and the bowed stone snapped in the cascade – giving Gavin, whose main responsibility was to document the project photographically, the closest shave of his career to date (Fig. 6.21).

FIG. **6.18** The arrow

FIG. 6.19 (a) Extraction. (b) "Those are nice!" (Alexander Thom). (c) The bowed stone extracted

At our next meeting with Prof. Thom in Dunlop, he reiterated his opposition to the project. But he couldn't restrain his reaction to the photographs: "Oh, those are nice," he said, as he reached Fig. 6.19b, and he mellowed sufficiently to give his advice on how they should be placed (Fig. 6.22). His first reaction to Bill Jones's plan was critical. Stabbing at the major standstill alignments with the stem of his pipe, he demanded, "I hope they're not going to be opposite? – Because they're not, you know!" I reassured him that both parallax and refraction would be taken into account, once we had a proper survey of the alignments (Chap. 7; Fig. 6.23). Mollified, he went on to tell us that the largest stones should be allocated to the lunar sight-lines, because that's how it was done at the ancient sites.

Using the largest stone of all as a table, and armed with that information, John and I found it surprisingly straightforward to assign each stone to a place (Fig. 6.24). But by that time the stones were already at the foot of the Broomhill, on the Pinkston road-side. Events had begun to move very fast indeed.

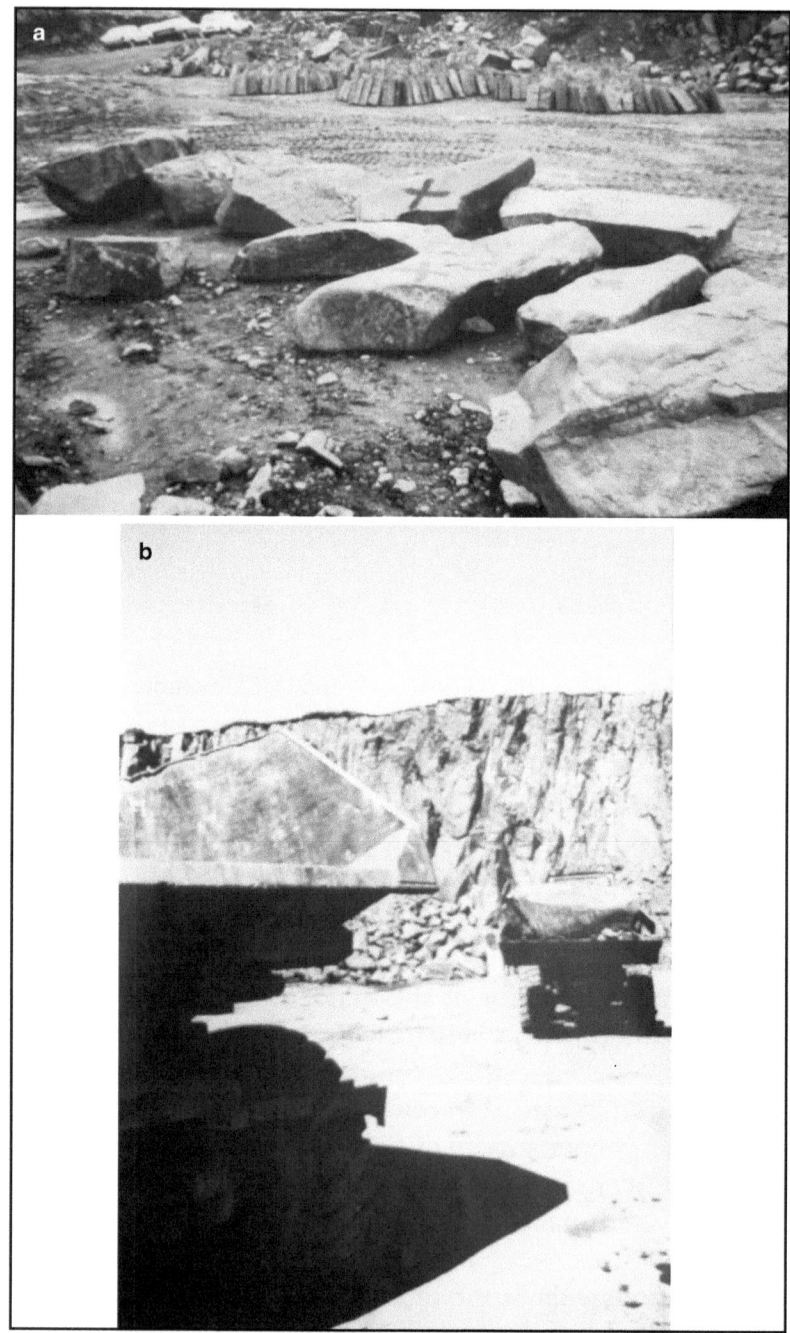

FIG. 6.20 (a) Stones lined up for uplift; the quarry's regular output at rear. (b) The first stone uplifted. (c) The first stone into the dumper. (d) The bowed stone goes into truck no. 2

FIG. **6.20** (Continued)

Fɪɢ. **6.21** The end of the bowed stone (in the air, at *left*, coming towards the camera). Notice the lack of shake. Gavin said, "If it was to be the last photo of my life, I wanted to make it a good one"

FIG. **6.22** Conference at Dunlop (Dr. Archie Thom at rear)

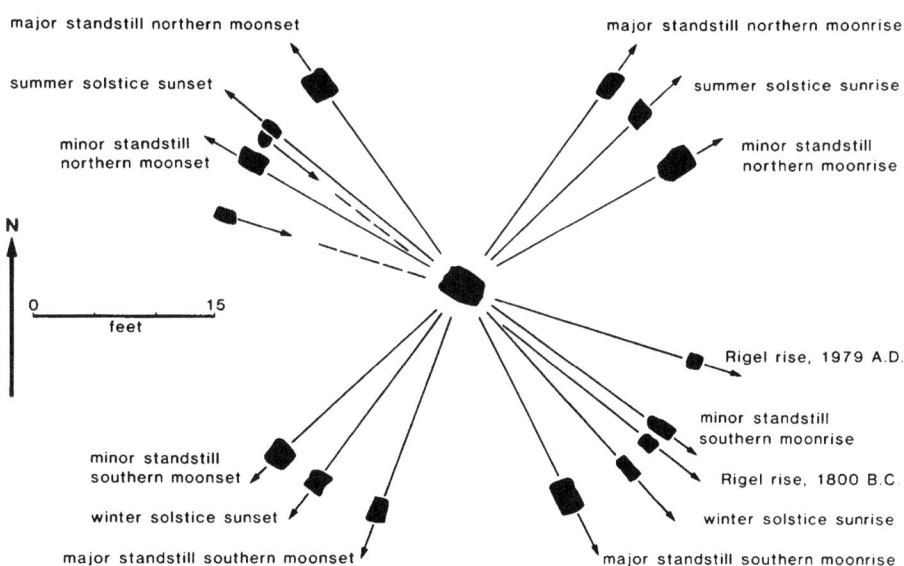

SIGHTHILL, A MODERN MEGALITH
GLASGOW
ASTRONOMICAL ALIGNMENTS

FIG. **6.23** Sighthill circle alignments, corrected for parallax and refraction (Drawing by Dave McClymont)

Fig. 6.24 Organizing the stones

References

1. Miller, H.B.: The History of Cumbernauld and Kilsyth. Cumbernauld Historical Society, Cumbernauld (1980)

7. Operation Megalithic Lift

You know, Hawke, this thing has caught the Navy's imagination *exactly*.

> – Sydney Jordan and Willie Patterson, 'Overland,'
> *Jeff Hawke, Daily Express*, 1966–1967 [1].

The theoretical alignments still had to be modified to the precise contours of the skyline. Although by Thom's standards my design was very imprecise, with each marker stone occupying several degrees of the field of view from the other side of the circle and no use of distant outliers or horizon features as 'foresights,' small differences in the elevation of the actual horizon would translate into much bigger differences in azimuth. To pin down the solar and lunar alignments with the accuracy found in the prehistoric sides must have required years of patient observation (decades at least for the lunar standstills), using prominent features on the distant horizon as 'foresights,' with teams of observers spread across the landscape to pin down the exact view station with greater and greater accuracy. To find the most northerly and southerly positions of the Moon, rising and setting, the ancient astronomers must have refined their sight-lines over a century or more before commemorating them in stone.

There was no question of using such methods at Sighthill, where the site was already fixed and the timescale was only months. But finding the actual rising and setting positions was difficult. A lunar standstill had occurred not long before the start of the project, so there were 9 years to wait for the next one. The 1978 midsummer solstice had passed before the site was finalized, and the weather around the winter solstice was atrocious, so no observations had been possible at the point when the pressure to finalize the design came to a head.

Although we knew the answers, in theory, fitting them to the actual skyline, allowing for parallax and refraction, was a different

D. Lunan, *The Stones and the Stars: Building Scotland's Newest Megalith*,   189
Astronomers' Universe, DOI 10.1007/978-1-4614-5354-3_7,
© Springer Science+Business Media New York 2013

FIG. 7.1 An unsuccessful survey. Note the bush in the nor-nor-west, bending over in the wind

matter. We couldn't hope to run the calculation backwards and match the accuracy of the prehistoric sites, without foresights such as notches or peaks on the distant horizon. But with good enough bearings taken, we could solve the problem graphically to within a degree.

John Braithwaite had arranged the use of a theodolite – old, temperamental, but still good enough for the standard of accuracy we acquired. In October 1978, however, soon after the stones had been deposited at Ruchill, the weather took a major turn for the worse. Even on the days when the skyline was clear, high winds made the theodolite virtually unusable (Fig. 7.1), and the wind chill fogged our brains so that when we did get a set of bearings and retreated to the office to warm up, the results was so inconsistent as to be meaningless.

After our first two unsuccessful sessions we called on the Parks "1,001 Project" survey team for help, and by January we had made several unsuccessful visits to the hilltop with them. During this time our respect for Thom's achievements increased a great deal. The site was still waste ground, and word spread around the Parks Department that you could tell when the Astronomy Project had been to the hilltop (and afterwards, whether we'd been to the toilet,

FIG. 7.2 True north bearing in the Campsie Hills, from the viewpoint of the future center stone

or gone to the kitchen for cups of coffee) by the trails of mud that we left through the building. I had inherited a number of high-quality, durable pairs of brogues from my late uncle Gordon, but by the end of the project, I had a great deal fewer that were still wearable.

One particular problem was the skyline in the east, look-ing out of the city and on through the haze of the steelworks in Motherwell. In the original panorama a line of hills clearly showed beyond – or was it just a line of cloud? At last on February 23 we had a miraculously clear, still, reasonably warm day when the sur-vey team was available and the survey was completed.

By then we had decided that the most effective way to finalize the layout of the circle was with big photographic blowups of the skyline, large enough for grids to be drawn 2° on a side, in the center of each print, without major distortion. Gavin had repho-tographed each alignment, using his panorama and the magnetic bearings as guidelines. The first of these 'working photographs' was of the view due north to the Campsie Hills, where the true north point could be identified from the Ordnance Survey map (Fig. 7.2). On each of the other photographs, the theoretical align-

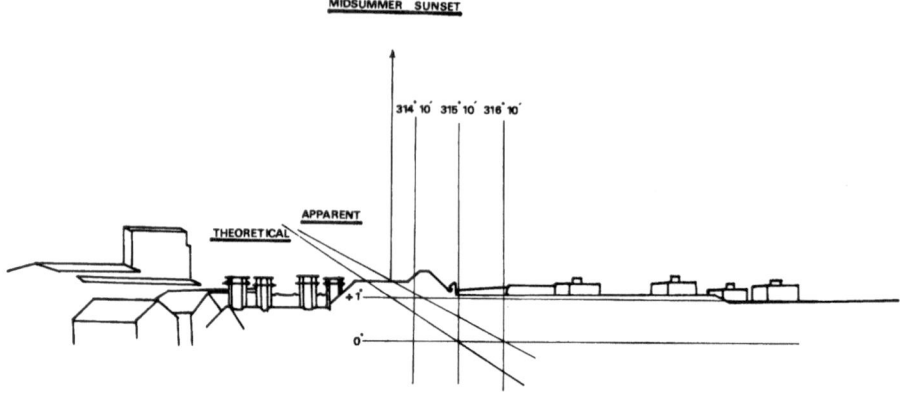

FIG. 7.3 Midsummer sunset as an example of the graphical calculation

ment to the skyline event was marked (this time, with parallax taken into account), and likewise the true horizon. Then the bearings one degree east and west were similarly marked.

So in each 'working photograph' I now had a small central grid (to minimize distortion) 2° on a side, centered on the theoretical alignment. On each grid I then plotted the theoretical rising or setting of the heavenly body concerned, and where it crossed the 0° and +1° lines I corrected upwards for the mean values of atmospheric refraction given by Thom in *Megalithic Sites in Britain*.

The last step was then to draw the apparent rising or setting path and where that line cut the skyline in the photograph, this was the azimuth on which the rise or set would occur, and therefore the bearing on which the marker stone should be placed. Of course running the calculations backwards in this way produced far less accurate results than the Thoms' surveys of prehistoric sites, but since our circle would be only 40 ft across and we were not working to distant foresights, I hoped the shortcuts were permissible. As at Kintraw, where the stone did not break the skyline as seen from the hillside view station, each of our stones would be only a pointer to the celestial event.

The simplified diagram of Fig. 7.3 illustrates the principle. Fascinatingly, Gavin Roberts recognized a similar pattern of scratches

Fig. 7.4 A Bronze Age 'working photograph'? (Ludovic McLellan Mann, "Ancient Sculpturings", William Hodge & Co., 1915)

on a slate from a Bronze Age tomb at Portpatrick in Galloway (Fig. 7.4). Could this be a Neolithic astronomer's equivalent of a 'working photograph'? Archie Thom became involved in controversy over the famous 'gold lozenge,' discovered in 1808 in Bush Barrow, near Stonehenge, which appeared to be marked up with astronomical observations (Fig. 7.5a). He and other archaeoastronomers were enraged when the British Museum polished it flat [2]!

MacKie has drawn attention to the 'Nebra Disc,' a bronze disc with gold leaf ornament, from Mittelberg in Germany and dated to approximately 1600 B. C., which he believes to contain similar information (Fig. 7.5c). The similarity could arise because the two sites are on nearly the same latitude, just north of 51° (Fig. 7.5d). But when that latitude gives the main solar and lunar alignments the form of a rectangle, perhaps it also tells us that the culture of astronomer-priests was pan-European [3]. One of the kerbstones at Knowth bears what appears to be a complete 'Thom calendar' for the year (Fig. 7.5e), and elements of the design are found at other Irish sites.

The February 1979 break in the weather came just in time, for in the meantime we had been tackling the problem of delivering the stones to the hilltop. At the start of the project Sighthill was still completely waste ground, with no vehicle access, and a heli-

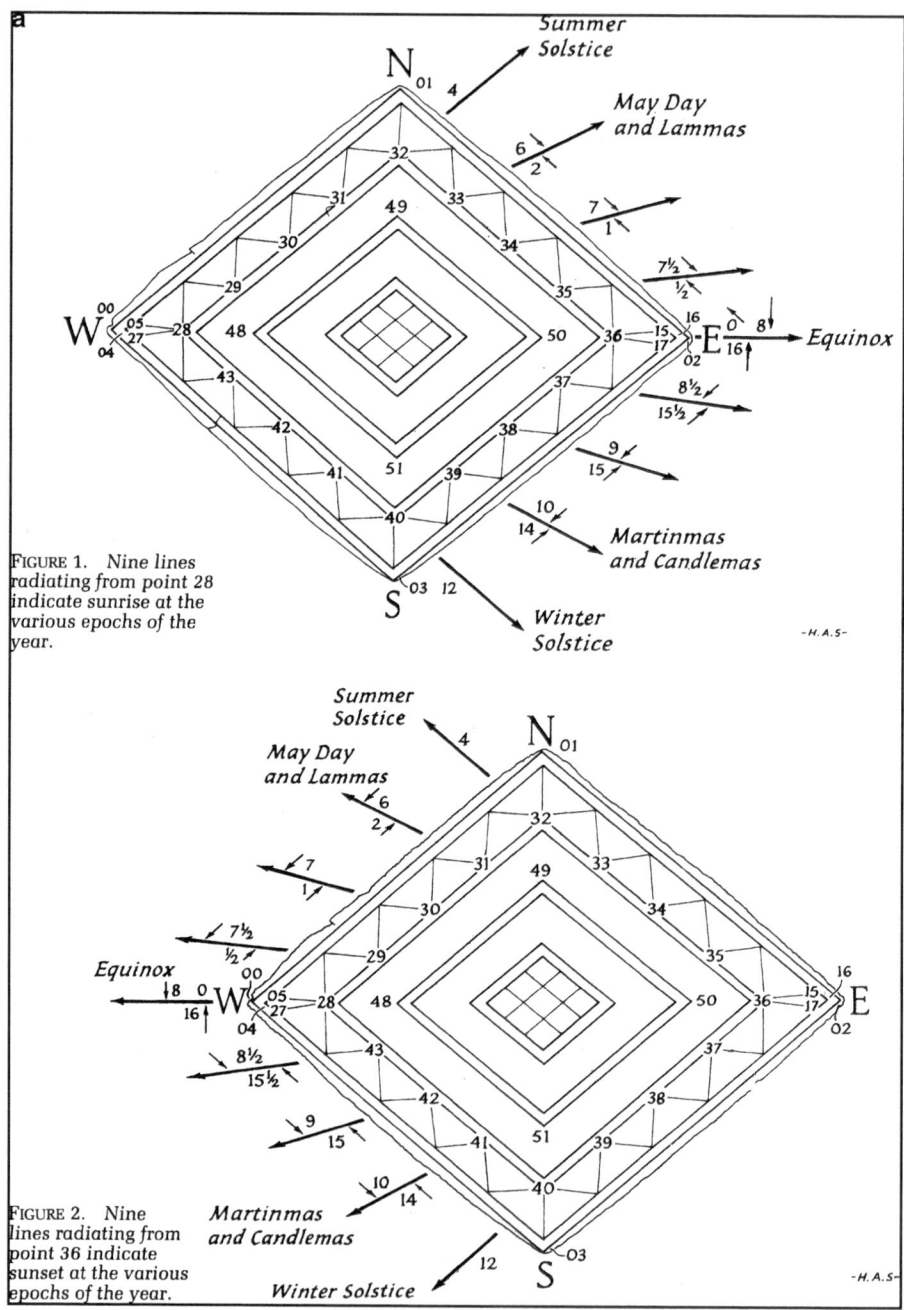

FIGURE 1. Nine lines radiating from point 28 indicate sunrise at the various epochs of the year.

FIGURE 2. Nine lines radiating from point 36 indicate sunset at the various epochs of the year.

FIG. 7.5 (a) Solar alignments on the Wiltshire lozenge (Supplied by A.S. Thom). (b) Lunar alignments on the Wiltshire Lozenge (Supplied by A.S. Thom). (c) Comparison of the Nebra disc and Wiltshire lozenge (note the

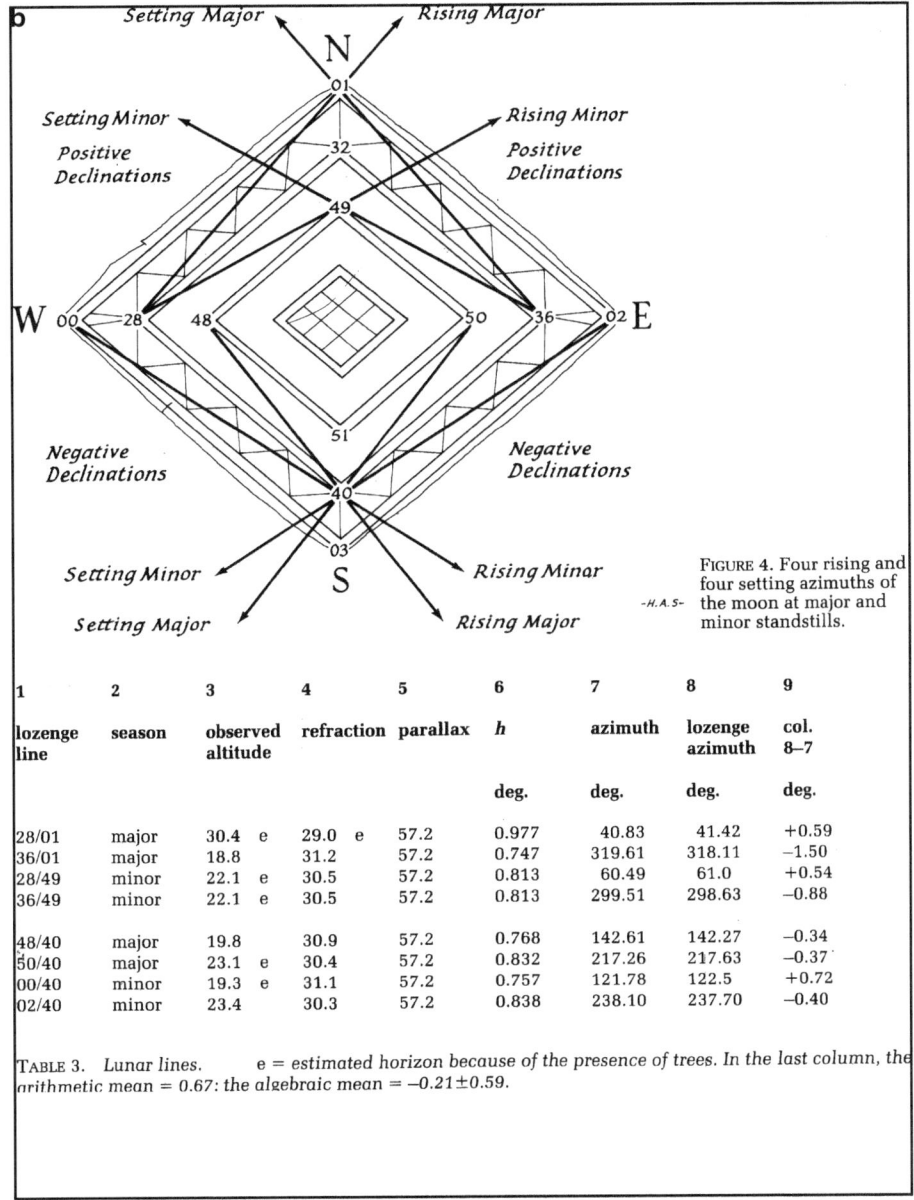

FIGURE 4. Four rising and four setting azimuths of the moon at major and minor standstills.

1	2	3	4	5	6	7	8	9
lozenge line	season	observed altitude	refraction	parallax	h	azimuth	lozenge azimuth	col. 8–7
					deg.	deg.	deg.	deg.
28/01	major	30.4 e	29.0 e	57.2	0.977	40.83	41.42	+0.59
36/01	major	18.8	31.2	57.2	0.747	319.61	318.11	−1.50
28/49	minor	22.1 e	30.5	57.2	0.813	60.49	61.0	+0.54
36/49	minor	22.1 e	30.5	57.2	0.813	299.51	298.63	−0.88
48/40	major	19.8	30.9	57.2	0.768	142.61	142.27	−0.34
50/40	major	23.1 e	30.4	57.2	0.832	217.26	217.63	−0.37
00/40	minor	19.3 e	31.1	57.2	0.757	121.78	122.5	+0.72
02/40	minor	23.4	30.3	57.2	0.838	238.10	237.70	−0.40

TABLE 3.   Lunar lines.   e = estimated horizon because of the presence of trees. In the last column, the arithmetic mean = 0.67; the algebraic mean = −0.21±0.59.

FIG. 7.5 (continued) Celtic calendar dates at *right*) (Supplied by Euan MacKie). (**d**) Latitudes of Stonehenge and Mittelberg (Supplied by Euan MacKie). (**e**) Possible 'Thom calendar' on Kerbstone K15, Knowth

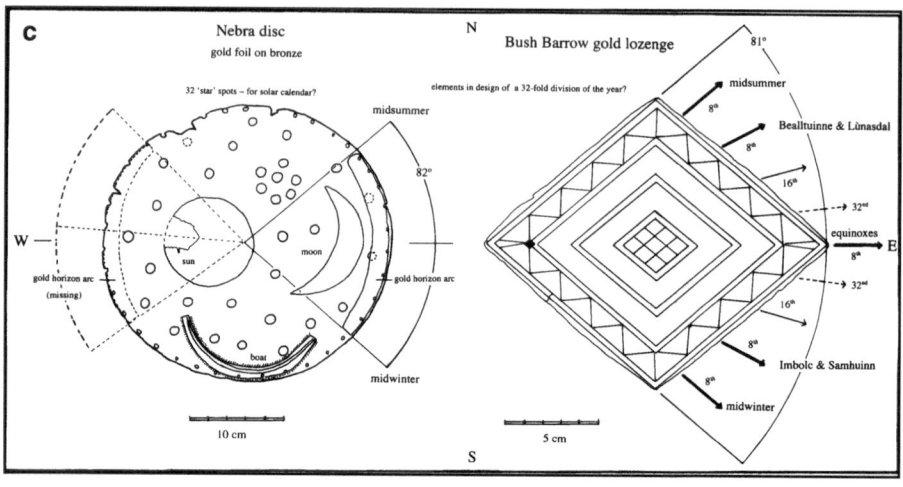

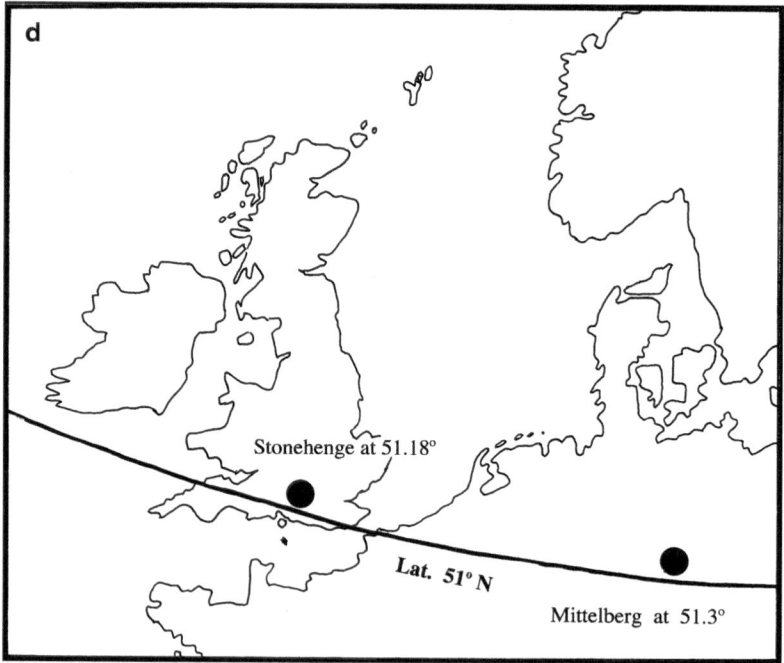

FIG. 7.5 (continued)

FIG. 7.5 (continued)

copter lift seemed the obvious answer. There was some discussion of a "living in the past" operation, using a large team of people to haul the stones onto and across the hilltop, but MSC Liaison and the Safety Officer were agreed that this was not an acceptable approach. Using a helicopter was in any case more true to the spirit of the project, in which the ancient principles were being adapted to the modern situation. Stonehenge and its counterparts had been built with the highest technology available, and to be faithful to those ideals we should do the same.

With the help of David Proffitt, I approached the Chief Pilot of the Navy's Air-Sea Rescue 819 Squadron at Prestwick, proposing to fly the stones to site by Sea King helicopter. The Navy was keen to do it, partly for the challenge, but also as a rare opportunity to conduct 'a weight-training exercise in urban conditions,' which wouldn't normally exist. The Ministry of Defense opposed it, but at that point John Braithwaite and I had a meeting with my boss, Ken Naylor. We hadn't met since he interviewed and hired John, and he began by apologizing because he hadn't even had time to read the report from me that he had requested. Then he picked it up, and his eyes bulged. "My God! Naval explosives experts – helicopters..."

"You told me to be self-motivated and get on with it," I pointed out.

"Oh, I did – I'm very pleased – I just hadn't quite expected this..." But he quickly rallied and focused on the MoD opposition to the helicopter plan. Within a week he had mobilized the support of multiple local councilors and members of Parliament, and although the MoD put obstacles in our path all the way to the event, suddenly the helicopter lift was on. We had to show that the job could not be done by a commercial company, but that wasn't hard. Apart from the cost of hiring a Sea King helicopter (£3,500 per hour), none could be spared from North Sea operations.

The Navy threw themselves into 'Operation Megalithic Lift' with a will; in fact, they gave us only 14 days to prepare for it. The commanding officer of HMS Gannet, Lieutenant-Commander Fraser Hutchinson, came to visit the project on March 2, and since the 819 Squadron had major defense commitments in the spring, and must always have a helicopter on standby for rescue operations, the megalith operation had to be carried out before the end of the month.

Setting up the meeting, Ken Naylor told me, "If you have any homework still to do for this, I want it on my desk tomorrow morning!" He didn't know what he was asking. Our successful session with the survey team had only just taken place, and I still had to transfer the results to fresh prints of the working photographs, let alone do the graphical calculations. Fortunately, by pure chance Tony Crerar from Wales (Chap. 4) was visiting me in Irvine at the time. He stayed up with me through the night, keeping me plied with cups of coffee and helpful suggestions, and I finished just in time to catch the 8.30 train and have the results on Ken's desk for 10 a.m. as specified. He had the grace to say, "I didn't realize how much work there was in this," to which I replied, "In that case, with your permission, I'm taking the rest of the day off and going to bed!" I used to pull 'all-nighters' regularly when I was a student, and from time to time as a writer, but at age 34, married, and in the middle of the working week, it was a different proposition.

We had known for some time that the normal lift capacity of the Sea King, fully fueled, was 3 metric tons, 6,000 lb. (Much

FIG. 7.6 (a) Uplifting the stones for weighing. (b) Delivering the weighed stones to Sighthill. (c) The big lunar stone (for major standstill south sunset) comes to Sighthill

heavier loads were to be carried during the Falklands conflict, 3 years later, but as David Proffitt told us on his return from it, few of the aircraft concerned were fit for operational use afterwards.) Arranging for the stones to be weighed proved remarkably difficult. To justify hiring tracked vehicles, dump trucks, etc., we were told we had to transport all the stones from Ruchill to Sighthill, and even then we were allowed to weigh only three of the stones as a sample (Fig. 7.6). I protested strongly against the extra work that would generate and the uncertainty it would introduce, but to no avail. Cost was the only consideration.

It meant that, knowing the weight of only three stones, I had to measure the maximum dimensions of all 22, calculate their weights as rectangular blocks, and compare the three known weights with the results. On average, they came out one-third lighter than rectangular blocks would be. Taking off one-quarter of each such weight for the other 19 stones, to be on the safe side, it became clear that some of the stones were much heavier than the quarry had estimated, because of the irregularities in their shape, and some would be too heavy for the Sea King to lift even with a light fuel load. The largest stone weighed over 4½ tons (Fig. 7.7). On that basis, six stones in all would be too heavy for the helicopter.

One useful check was that we had been given two figures for the density of whinstone: 114.24 lb/c.ft (2.9 g/c.cm, but we weren't using those on *this* project), or 123.9 lb/c.ft. The lower figure came out correct for the weights for the three stones that had been weighed, showing that my estimates of their linear dimensions were also very close. The higher figure couldn't be made to tally at all.

By October 1978 work had begun on Sighthill Park, and access was a great deal more easy. The helicopter operation could have been dropped, but symbolically and as a means of generating public interest it seemed too good to pass up. As Thom had told us that the largest stones should be used to mark the lunar alignments, for authenticity, we allocated seven stones to the helicopter: the solar stones, the contemporary star alignment, and the central stone. The lunar stones and the historical star alignment would be placed in position by Sportsworks, the contractors for the park, using JCBs, and would be cemented in place so that the foundations were ready for the helicopter's stones to be positioned.

FIG. 7.7 The major standstill stone – too heavy for helicopter lift

From the Navy's point of view, as a training and public relations exercise, the value of the operation was unchanged.

The foundations were discussed at great length with various experts after a preliminarily plan had been prepared by the Department of Architecture and Related Services and used in the application for planning approval. The prehistoric sites had no foundations, of course, although the pits for the stones were lined with branches or sticks to provide bracing, and stones placed on hilltops or sloping ground were carefully chosen so that the forces acting on the bases would tend to counteract slippage. Nevertheless many prehistoric stones have tilted or fallen in the last few thousand years, many of them, alas, with human assistance in recent times. It was felt such assistance might be all too forthcoming in the present, and the Planning Department modified the specification to have the concrete arcs beneath the stones pre-poured in reinforced concrete.

In the temporary set-up of the Jobs Creation and STEP programs, we had to work under or get round some particularly irksome restrictions. I've mentioned the business of putting purchases out to tender and the eventual crackdown on our use of petty cash. At the height of planning the High Frontier exhibition, when I was telephoning NASA, ESA and other national and international bodies on a regular basis, I was told for the first time (and chose to disregard) that the rules did not allow calls outside Glasgow, let alone overseas. Supposedly no official vehicle was to cross the city boundary for insurance reasons – a nonsensical ruling that frequently left people or equipment stranded on the wrong side of the line and led to Beltmoss Quarry bringing the stones into Glasgow with their own transport as a goodwill gesture. And because the Jobs Creation Scheme was so politically sensitive, all the Special Projects had to be conducted in secret.

In our case, the secrecy came to a sudden end when the Diary column of the *Glasgow Herald* published a reasonably accurate account of the project, with a cartoon by Turnbull (Fig. 7.8) [4]. Angry calls ensued from the Trongate headquarters of the Parks Department. I was science fiction critic of the paper at the time, and Christopher Small, the Book Page editor, had urged me to write a piece for the Diary, but permission had been refused. Of course I came under suspicion. But I had known nothing of the piece that appeared. Thinking over his movements in recent days, Gavin remembered a pub discussion with Bill Ramsay and other members of ASTRA. Going back to the bar, and thinking he recognized the eavesdropper, he accosted him with the words, "I think you owe me a pint," – and got one, with the reply, "Yes, I got a good bung from the *Herald.*"

Once the story was out, however, the Project had become a target. Some Scottish newspapers were campaigning against the Jobs Creation scheme on political grounds, and the *Daily Record* was seizing any stick with which to beat Special Projects in particular. They had a 'mole,' probably in the main Parks Dept. offices on the Trongate, who was passing on telex messages sent down from Buchanan Street. Since the telexes were passing between people who knew what they were about, very often the reporters got the facts wrong. Our friends at the *Record* usually spotted them and spiked the ones about us, but one they missed was inspired by the messages about the foundations.

Because the circle was to be on District Council land, it didn't require planning *permission*, only the less formal planning approval. But the story in the *Record* was all wrong, beginning, "It seemed like a great idea to put up a Druid's Circle [sic] for the benefit of amateur stargazers. But the organization was less than

"See that, 10 minutes late by the noon sighting!"

FIG. 7.8 Turnbull cartoon, *Herald Diary*, Nov 1978

magical. No one remembered to ask for planning permission, with the result that the whole project has been halted..." and ended with the inevitable quote from an anonymous Glasgow councillor who "summed it up best when he asked, 'How come the Druids managed it without planning permission or helicopters?'" [5].

After other projects had suffered similar derision John Braithwaite, who had started an Urban Legend for a bet when he was a student [6], decided it was time for a hoax. The Glasgow subway system, 'the biggest toy train set in the world,' was in a major upgrade at the time, with the biggest excavation just downhill from our office in Buchanan Street. John got together with other victims to start a rumor that a prehistoric fern seed had been found in the excavation and passed to Special Projects for rearing, only to be eaten by the office cat. Sadly, despite fake telex messages and lunchtime visits to the bar by the *Daily Record* to talk about the story, the paper failed to bite.

Once the foundation design had been modified, planning approval was granted without difficulty. There was brief confusion when it looked as if the stones had to be half-buried, but that was resolved with some sketches and measurements added by Gavin Roberts (Fig. 7.9).

However, now a bigger practical problem had arisen, regarding the lifting operation itself. Although elaborate slinging of the stones had been discussed, allowing the helicopter to hook on to

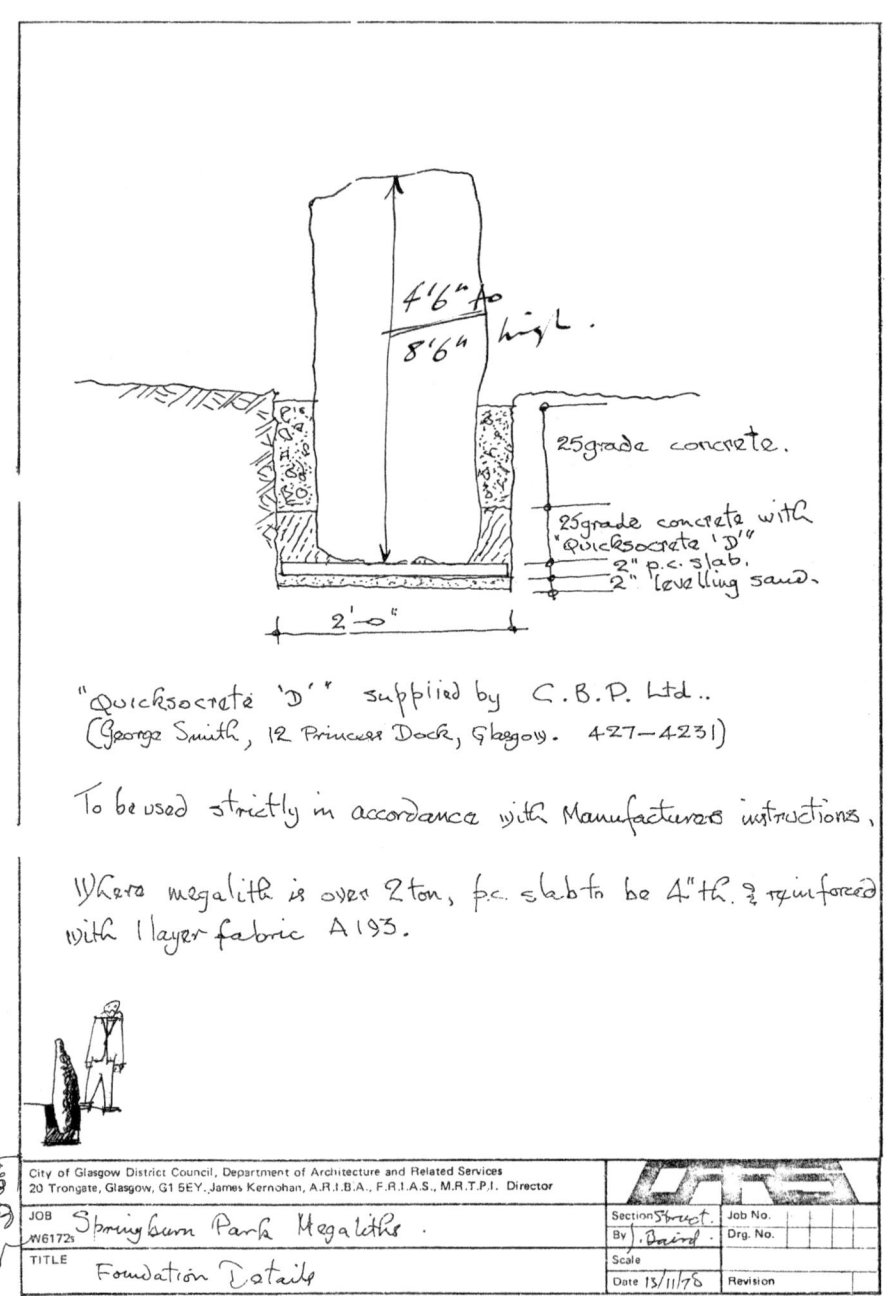

4'6" to 8'6" high.

25grade concrete.

25grade concrete with "Quicksocrete 'D'"
2" p.c. slab.
2" levelling sand.

2'-0"

"Quicksocrete 'D'" supplied by C.B.P. Ltd..
(George Smith, 12 Princess Dock, Glasgow. 427—4231)

To be used strictly in accordance with Manufacturers instructions,

Where megalith is over 2 ton, p.c. slab to be 4"th. & reinforced with 1 layer fabric A193.

City of Glasgow District Council, Department of Architecture and Related Services
20 Trongate, Glasgow, G1 5EY. James Kernohan, A.R.I.B.A., F.R.I.A.S., M.R.T.P.I. Director

JOB Springburn Park Megaliths.
/N6172s
TITLE Foundation Details

Section Struct.   Job No.
By J. Baird.   Drg. No.
Scale
Date 13/11/78   Revision

Fig. 7.9 Planning Department foundation specification (Amended by Gavin Roberts)

each stone and place it vertically with accuracy, it would take too long to fix them in position – apart from the general reluctance to have Sportsworks personnel working under the helicopter. It had been clear for some time that nets would be used, and the Sea King would simply lower each stone into the prepared pit and cast it off.

At first it was supposed that the nets would have to be sacrificed, with the lower parts buried in the cement and their tops cut off at ground level. But if Sportsworks men and machines were going to sling the stones and raise them upright in any case, they could be lifted high enough to clear the nets for retrieval. At first the Ministry of Defense was adamant that arranging suitable nets would be our responsibility, and John Braithwaite had contacted various shipping and cargo handling firms in the city, looking for suitable nets and ropes. 'A tip of the hat' is due to Martin Black's and to Nevisport, who were willing to provide nets and to sacrifice them if necessary, though they were relieved to hear that they would get them back.

With only days to go, however, the Ministry of Defense informed us that its own nets were to be used, and the supply of them from the Royal Air Force was being flown north to HMS Gannet over the weekend. Lieutenant Norman Leask, who was in charge of this part of the operation, was relieved to learn that at the end of it they would be returned intact. But whereas the nets we had arranged for were stressed to 10,000 lb weight, the RAF ones were stressed for only 5,000 lb. I estimated the weight of the central stone at 4,200 lb, but if I was off by just one inch in my estimate of its mean diameter the weight would rise to 4,800 lb. Norman Leask was unhappy with my calculations and did his own for the central stone, adding a safety factor of 25% instead of subtracting it as I had, and came up with a weight of 6,678 lb – too much for the helicopter, let alone the nets.

I pointed out that if his figure was correct, all the rest, including the weighbridge results, had to be wrong. It would mean that the central stone had 74% of the weight of the biggest lunar stone, although they were made of the same rock and it had less than a third of the volume. Still, with the new nets the safety reserve vanished. Should we abandon the symbolism of having the helicopter complete the structure with the central stone, after proceeding sunwise around the circle from the midsummer solstice? Sticking to the 'no publicity' mantra, the Director of the Parks Department, Keith Fraser, had opposed the helicopter operation, and if it had to happen, had urged that it would be enough to use the helicopter for just one stone, any stone, preferably at first light where nobody would see it; but if the aim was public spectacle, in tribute to the ancient builders, then Ken Naylor and I felt the helicopter should be used to build as much of the circle as possible.

Details of the big day were settled at a final meeting on March 16, chaired by Ian Clair and attended by Sportsworks personnel, Norman Leask from HMS Gannet and his colleague Lt. McBride, who dealt with the aircraft. There were also representatives of the Fire Service, Strathclyde Police and St Andrews Ambulance Brigade, John Braithwaite, Gavin Roberts and myself, and Robert Hastie, manager of Glasgow Parks Broomhill District, who had organized transport throughout the project and was now responsible for crowd control. There were still pressing points to be settled regarding insurance coverage for the operation, which did not come through until the very last moment. The helicopter was to be insured for £1.5 million, plus third-party insurance claims up to £1 million, with the crew insured for £110,000 per officer and £95,000 per rating, for the aircrew and for two teams of one officer and three ratings each, on the ground.

Refueling facilities were required for the helicopter, which would be lightly fueled for maximum lift. A nearby truck park could not be used because the owner of the company was on holiday in Majorca and was not to be disturbed. The Electricity Board was agreeable to the use of the waste ground where the power station had been demolished, and they, too, were represented at the meeting, but after inspecting the site the naval officers declared it unsuitable. Although the rubble was unlikely to damage the landing gear of the helicopter, the real danger was that some fragment would be picked up and later fall to cause injury or damage. However, nobody told the watchman on site, who came up to the circle to see whether the Electricity Board ground was to be used. Looking around as the lunar stones were being erected, he made the comment that led us to the historical Knappers site (Chap. 5).

Refueling facilities were therefore to be arranged at Abbotsinch Airport (now Glasgow Airport), and the helicopter would probably have to break off to refuel in the middle of the operation. The local schools, who had been invited to watch, had to be advised accordingly. Wellington boots and protective helmets had to be arranged for all personnel who would be on the site. Two press conferences, a reception, and a leaflet issued to the schools were all still in the planning stage.

The Fire Service representative asked if hoses should be run up the hillside in advance of the operation. The naval reply was that apart from the negative psychological effect that would have on the aircrew, it would be highly dangerous to the aircraft and the ground crew if the hoses began whipping about in the hurricane-force downdraught below the rotors.

"But if the machine falls out of the sky..." the fire officer began, having evidently picked up some naval phraseology.

"I think this is getting out of hand," said Lt. McBride. "We can lose a tail rotor blade and stay in the air. We can lose a main rotor blade and stay in the air. We can even lose one of the *engines* and stay in the air... and if anything is going to fall out of the sky, gentlemen, I regret to inform you that it is going to be the *rock*."

That was getting a little too close to home, and Lt. Leask and I exchanged meaningful looks. I had warned him that there was still a lot of opposition to the operation within the Parks Department and even, I had been led to believe, within the District Council. We were being bedeviled by spurious phone messages, coming from somewhere within Trongate or the City Chambers, trying to order essential personnel on to other tasks, or saying that the whole operation had been canceled. There had been no time for a rational discussion about the question of the central stone, and if it was reported to management at that stage, I felt sure that it would be used as an excuse to cancel not just the flight of the last stone but the entire operation.

Norman and I went around to the base of the hill, and we stood on the problem item. There was just room for both of us. "To be honest, it doesn't look that big," said Norman.

"I don't think it is that big," I replied. "I've gone over the calculations repeatedly, and I don't see how it can be too big."

"If you're completely confident of your figures, then we'll fly it," said Norman. "But I have to make this clear: if anything goes wrong, it will be entirely your personal responsibility, not ours."

It needed a moment's thought. But when we had come so far, and I *was* confident, I took the plunge and said, "All right, let's do it." Ken Naylor's initial requirement that I be self-motivated and work without direction had come home to roost in a way that neither he nor I had anticipated. I wonder if I would have the nerve to do it today.

Trenches had now been prepared in four arcs to surround the central stone, each lined at the base with reinforced concrete. The lunar stones were to be slung from the blades of two earthmoving machines, placed on the reinforced concrete base of the trench, and aligned using the 'working photographs' to match up their positions. Each stone would then be wedged in place with beams and propped up on pieces of brick about the base, before boards were placed across the foundation trench and quick drying cement was poured to bury it by up to one third of its height. The operation was originally scheduled for March 18, and I drove up from Irvine to Glasgow through a blinding snowstorm. Jimmy, the Sportsworks foreman, gallantly came to the site office to meet me, but it was obvious that the operation had to be postponed.

The next day the weather was fine, with thin snow on the muddy ground, though the air grew misty later. The Sportsworks operation on the hilltop began early. Jimmy explained that one JCB would be active, one passive, as the stones were maneuvered. "But don't worry, this guy [driving the active machine] is so good he could load and unload eggs." The first stones to be erected were the northern moonrise ones for the major and minor standstills, flanking the space for the midsummer sunrise stone, under my supervision. It was a thrill to see the first stones upright, so many months after we had chosen them. Joining me as we moved from the northeast arc to the southeast, John Braithwaite was quite emotional: "Christ, it's beautiful. I never thought it would look this good." Gavin Roberts then took over, working sunwise around the circle, leaving the gaps into which the helicopter would place the remaining stones, while I went to a press conference at the City Chambers, also attended by Fraser, Naylor, Dr. Roy, Archie Thom, and Councillor Robert Logan, Chairman of Glasgow District Council's Civic Amenities Committee. Roy said it was appropriate that this tribute to the early astronomers should be paid just when the *Voyager 1* spacecraft was reaching Jupiter as our most advanced probe into the universe to

Fig. 7.10 Minor standstill southern moonset stone, with M8 motorway behind

date. Dr. Thom said that hopefully the project would draw attention to the need to preserve the ancient sites as a cultural heritage.

Interruptions during the day continued without respite. Three times, I had to break off and go over the hill to the Sportsworks site office, because a messenger had arrived telling the crew to stop work – 'person or persons unknown' had called to say the helicopter operation had been canceled. If they *had* stopped work, it would have had to be cancelled because the foundations would not have been ready. Each time, I asked Jimmy and his team to continue, handed over to Gavin, went to the telephone and wasted an hour confirming that the operation was still on.

When I got back to the site the southwest arc was done (Fig. 7.10), and I took over for the final northwest one. For some reasons those stones were particularly difficult to place, possibly

FIG. 7.11 The last Sportsworks stones in place

FIG. 7.12 Preparing the foundation for the central stone. Author with working photographs underarm, Royal Infirmary and Glasgow Cathedral at rear, beyond the M8 motorway

because we were all tiring. One of them took the concerted efforts of all present to hold it on to the alignment, and it insisted on rotating before it could be wedged upright, so the northwest arc of stone is not as tidy as I would have liked. But with them all in place (Fig. 7.11), the foundation for the central stone could be prepared (Fig. 7.12), and we could call it a day at the site.

On the public relations side, due to the short notice, there were still major commitments to be fulfilled. In late 1978 plans had been drawn up for a much enlarged educational astronomy project, and we now had an illustrator, David McClymont, who was working on a leaflet to be handed out to the schoolchildren and the public during the operation. We were taking part in an ambitious class project at Faifley Primary School and Ian Downie, the Astronomy Section head of ASTRA, was standing in for us on that and other commitments. Now we were asked to prepare a display board for the press conference at St. Stephen's Primary School the following morning, March 20. John Braithwaite stayed at the office overnight to do so, and Dave McClymont mounted it up immediately beforehand, while hundreds of copies of his leaflets were being run off. Other ASTRA members who had come in to help were pressed into service folding them, while spaceflight display boards on loan to the project from ASTRA were mounted in the S.T.E.P. building recreation room in preparation for the reception after the event.

One of the trials of Special Projects was an individual who was known to those he shared an office with as 'The Vicar' – a designation that particularly annoyed him, since he was fervently Presbyterian. His project had no other members, from which you may draw your own conclusions. We called him 'Ezekiel' because the astronomical or extraterrestrial content of that Old Testament book was his normal conversational gambit. He used different ones with other people, but whatever the opening, within 4 min he would have reached his serious business, a personal message that began, "You see, I believe that Jesus is our personal savior," and go on in that vein for as long as he was allowed to. My personal defense was to interrupt regularly, saying, "Yes, but that's just what *you* believe," until he lost his temper and gave me an excuse to end the attempted conversion. It could not be called a conversation or a discussion by any normal social definition. John's reply was more direct, namely that the next time he interrupted work on the Astronomy Project his faith would be put to the test by ejection from our third floor window.

That worked until the morning of helicopter day, when the urge to preach became too much for him. As he entered the office after we and the ASTRA volunteers had spent 2–3 h folding leaflets for the schoolchildren, he blocked my first wife's exit with the cry, "I trust there is nothing mystical about this operation..."

"It's far too cold up there to be mystical," she said, "get out of my way!" We were in a serious hurry to get to the press conference at the school nearest to the site, and things were starting to go awry. He was left protesting to empty air, "That's not what I meant...," and was not the only person to be left in that situation as the day continued.

The press conference at the school was for the signing by the Thoms and Mackie of copies of their books to be placed in a time capsule beside the central stone, along with press coverage of the event and a message from the Lord Provost David Hodge. In the end the plastic capsule was not ready and a surrogate had to be used. Mackie was unable to attend the signing ceremony, and the torch to seal the capsule was broken. The message from the Lord Provost did not arrive until later, so the placing of a lunch box to represent the capsule for the benefit of the press was purely symbolic, though like President Nixon's speech to the astronauts on the Moon, it used up a lot of very valuable time.

John had already been on site with the Sportsworks JCB team and the naval ground crew, getting the stones into the RAF nets ready for the helicopter before coming over to the school (Fig. 7.13). The rest of us arrived at the same time as the Thoms, as hundreds of schoolchildren began to gather at the rallying point. Getting out of their Range Rover, at first the professor didn't recognize me, and Archie Thom had to reintroduce us. "I'm still against all of this," he reminded me. There were strict instructions that nothing was to happen until the Lord Provost's message arrived, and as we waited helplessly, the Sea King appeared overhead precisely on schedule.

John left with the naval ground crew and I followed as soon as I could, with the helicopter already spiraling in. It was agreed that my wife would go with the Thoms in the Range Rover, to watch from the path around the base of the hill, surrounded by cheering children. Initially Gavin was with John (Fig. 7.14a), alternating again between

Fig. 7.13 (**a**) John with Norman Leask and ground crew. (**b**) Slinging the stones – Jimmy directs. (**c**) Stones slung for lifting

c

FIG. 7.13 (continued)

color and black-and-white cameras, and as John told me afterwards, "He has some bottle as a photographer. He was getting right under the chopper and the stone for some of his shots (Fig. 7.15d), and I was having to brace him with all my strength against the downwash" (Fig. 7.16). After seeing the first stone lifted (Fig. 7.14, 7.15, and 7.16) I made for the hilltop (Fig. 7.17a). I was supposed to be alone up there except for the other naval contingent, the Sportsworks team, and the film crews from the BBC and Scottish Television (Fig. 7.17b). As the

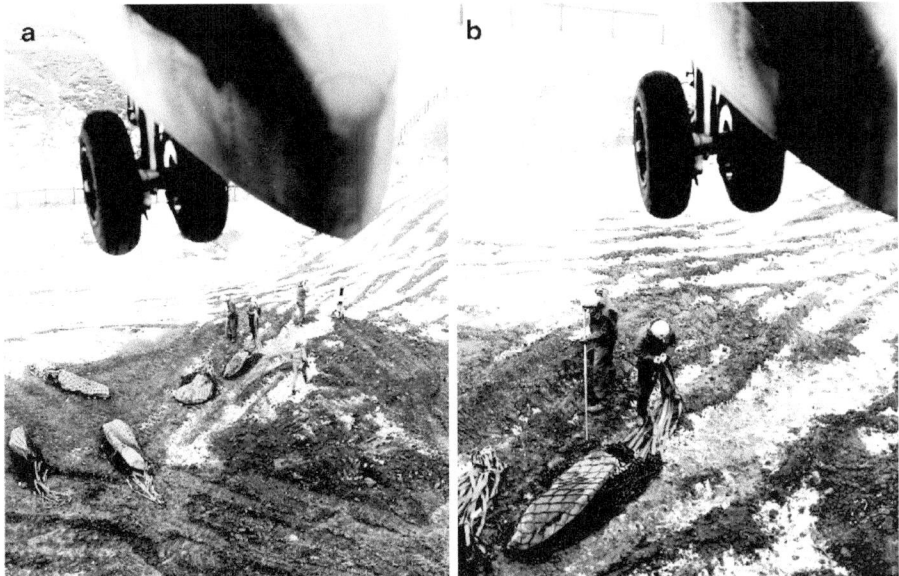

FIG. 7.14 (a) On approach. Central stone at *lower left center*, Gavin Roberts at *right center*, John Braithwaite at *upper right*. All Royal Navy photographs by Leading Photographer Burnie. (b) Readying the first stone (to mark midsummer sunrise)

operation went on, however, more and more people talked their way through the police cordon and joined me up there, as can be seen in the photos taken from the helicopter. One of the first to join me was Archie Thom, who made it up and down the mud and snow of the steep slope despite being in the kilt, of which he was fond. There was no safety problem, since anyone who felt the Sea King's 100 mph downdraught kept well clear thereafter. Each time the helicopter withdrew, however, a small crowd swarmed over the hilltop to see what had been added to the circle, only to scatter as the rotor blades reappeared over the edge of the hill.

The day was cloudy, with a stiff cold wind that actually helped, since it gave the helicopter greater lift. The school parties and other onlookers made a crowd of at least 1,000 on the footpath bordering the park above the M8. The police had decided against closing the motorway, but had difficulty in keeping the traffic moving. One fascinated motorist was apprehended trying

FIG. 7.15 (a) Hooking on – Sighthill high flats at rear. (b) Taking the strain. Note the 819 Squadron badge, commemorating the Swordfish raid on the Italian fleet at Taranto in November 1940. (c) Lifting the first stone. (d) The midsummer sunrise stone airborne

to reverse back up in the fast lane. As Alexander Thom watched from the footpath, from where he had a good view without leaving the car, the six stones were placed around the perimeter without trouble (Figs. 7.14, 7.15, 7.16, 7.17, 7.18, 7.19, 7.20, and 7.21), and keeping accurately to Lt. Leask's predicted time of 5 min per stone. Every so often there was a slightly unnerving pause as the pilot conferred with the ground crew; they had no radio link, and he had to drop a headset on a cable, while the helicopter hovered on auto-pilot, nodding and wobbling in what seemed dangerous proximity to the ground. From the last of those conferences, however, came

FIG. 7.16 The first stone in flight. Note John bracing Gavin at *lower left center*. author at *lower right center*, before scaling the fence on the way to the hilltop (Photo by Frank O'Neill)

good news. Fuel consumption was lower than expected, and the job could be finished without a break for refueling at the airport.

By that time Archie Thom felt he should rejoin "the old man," slithering down the muddy hillside in his kilt; and with my tension mounting as I waited for the last stone, suddenly I was joined on the hilltop by the last person I expected: Keith Fraser.

"It's going very well, Duncan," he said.

"Yes, so far," I replied. I intended to say no more, but the tension in my voice gave me away.

"Duncan, is there something you should have told me?" he asked, and being in no mood to dissemble, I told him succinctly how the weight of the central stone was the only unresolved issue. As we spoke we could hear the regular beat of the Sea King's rotor from out of sight at the northeastern base of the hill. Just as Keith opened his mouth to respond, the beat changed steeply, rising to a whine. As the rotor blades came into view it was clear that they were now at a much steeper angle, and as the fuselage followed, it was oriented straight into the wind, instead of inclined to it as it had been hitherto.

FIG. 7.17 (a) First approach to the hilltop. Royal Navy photograph. (b) BBC and Scottish Television crews on site (Photo by Frank O'Neill)

In retrospect, the problem could be seen from the outset, in Fig. 7.14a, and clearly in Figs. 7.22a, b. Due to an error by the ground crew, the stone had accidentally been slung with the heavy end uphill, and as it lifted it began to slide, dragging down the nose of the machine and creating a risk that the blade tips would strike the ground. John told me afterwards, "I first realized something was wrong when I found I was alone – the ground crew had taken to the scenery." The pilot immediately increased the pitch of the

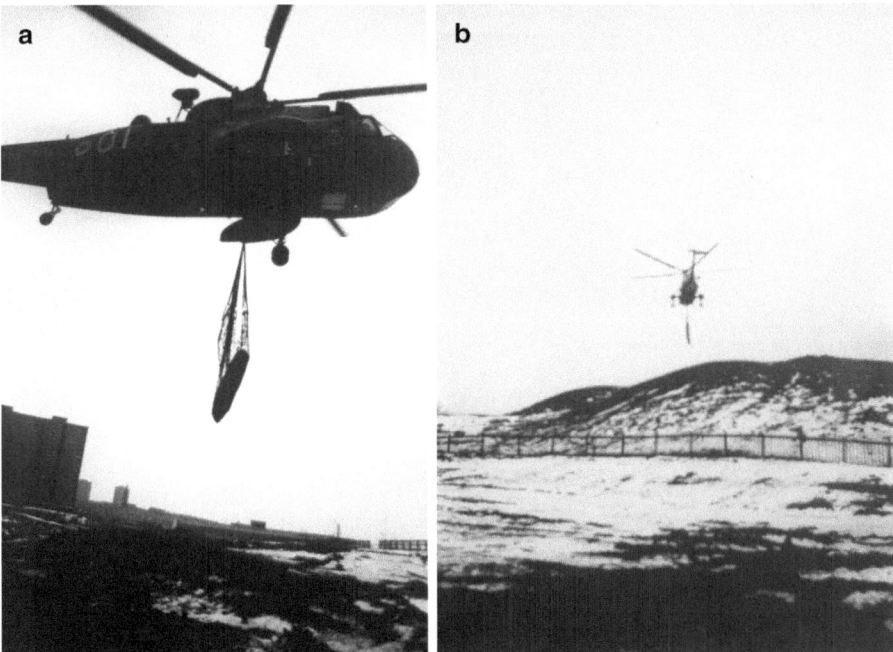

FIG. 7.18 (a) Second stone (Rigel rising alignment). (b) Rigel stone in flight

rotor blades and stepped up the power (giving those of us out of sight over the hill a nasty moment) and lifted the stone without further trouble, although this time he flew straight into the wind instead of angling the helicopter to it.

"Duncan…" Keith Fraser began menacingly, but I was not in the mood.

"It's in the air, Keith!" I interrupted, "and it only has 200 yards to go. There is no point in worrying about it now." Sure enough, the net held during the traverse of the hilltop, and the megalith's main structure was completed (Fig. 7.23). With the help of the Royal Navy, we had given a ring of truth to the legends that stones were transported to the megalith sites through the air.

In 1998, Norman Leask was to be the pilot of a helicopter that saved ten seamen from a Bahamian registered vessel sinking off Shetland, and gained an MBE. (His winchman, who was washed overboard and died, got the George Medal). But somewhere in the final photos taken from the Sea King during 'Operation Megalithic

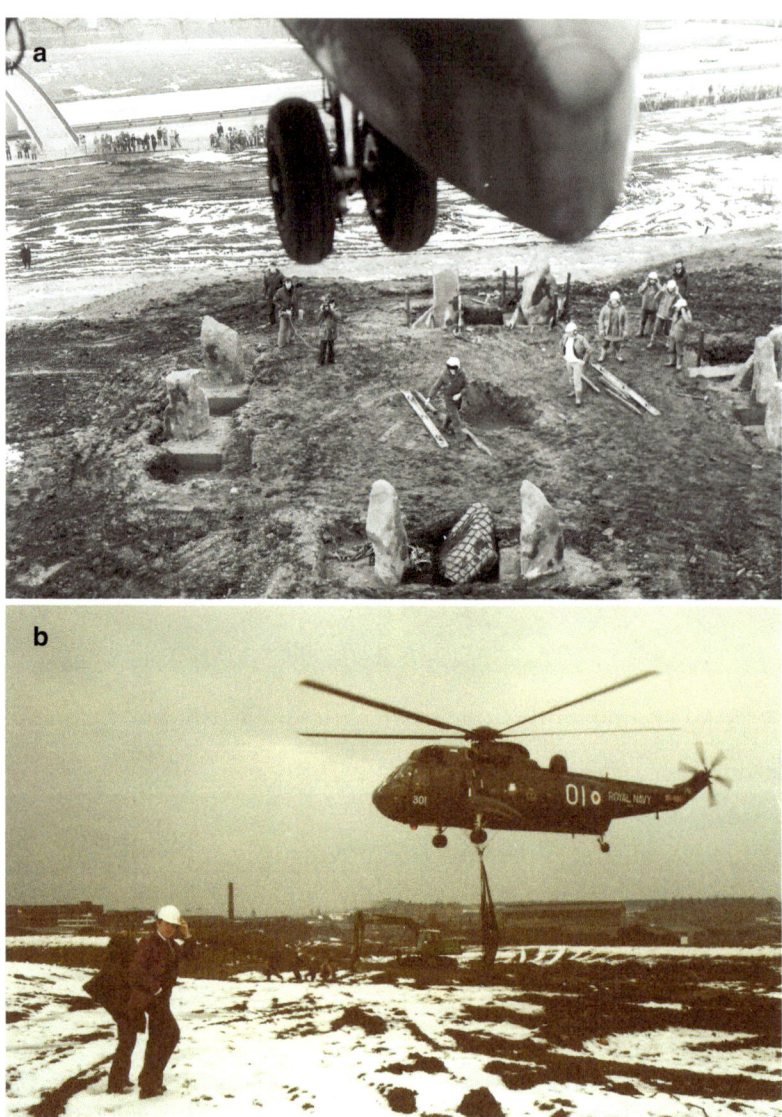

FIG. 7.19 (a) On approach with the Rigel stone; note the crowd on the footpath. Royal Navy photograph. (b) Rigel stone incoming (Photo by Frank O'Neill)

FIG. 7.20 Midwinter sunset stone, incoming. Note the helicopter's '01' designation

Lift' (Fig. 7.24), Norman is with Jimmy, the foreman of the park contractors, Sportsworks. Unknown to me, Jimmy had bet him that the Navy couldn't fly the seven stones in 35 min, and a large sum of money was passing from him to Norman now that they had.

After a short speech by Councilor Logan, Scottish Television conducted an equally short interview with me – it was very cold up there by that time. The BBC then conducted a longer one, in which unfortunately they ran out of film before the last question, which they considered very important, about why building the circle was worthwhile. I had to hold my position for what seemed like an eternity while they went for more film and reloaded, and, not surprisingly, people told me afterwards that my last reply lacked the conviction of the earlier ones. But it was good that everyone else had gone to the reception, and I was left to walk back to Buchanan Street alone. I really needed a few minutes to take in what had been achieved.

There was a pleasant development at the reception. We had now been joined on the project by John Braithwaite's father Bill, late

FIG. **7.21** Midwinter sunrise stone, incoming

of Charles Frank Limited. It turned out that he was an old friend of Thom's, and they greatly enjoyed their reunion (Fig. 7.25). Keith Fraser proposed a toast to the professor, and all present joined with enthusiasm.

I had been expecting Lieutenant-Commander Fraser Hutchinson to drop out of the helicopter before it left for refueling, but to my surprise he was already at the reception. As the Sea King was about to take off at Prestwick, a car had turned in at the gate and an official had stepped out, saying, "Ministry of Defense, snap inspection." Without hesitation Fraser had replied, "Very interesting

FIG. 7.22 (a) Approach to pick up the central stone; note the orientation to the fence uphill. (b) Picking up the central stone. (c) Final approach to the circle with the central stone. College of Building and Printing at *left*, St. Andrew House at *right*. Note the larger crowd on the hilltop. Royal Navy photographs

FIG. 7.23 (a) The central stone, incoming. (b) The central stone on finals. Royal Navy photograph

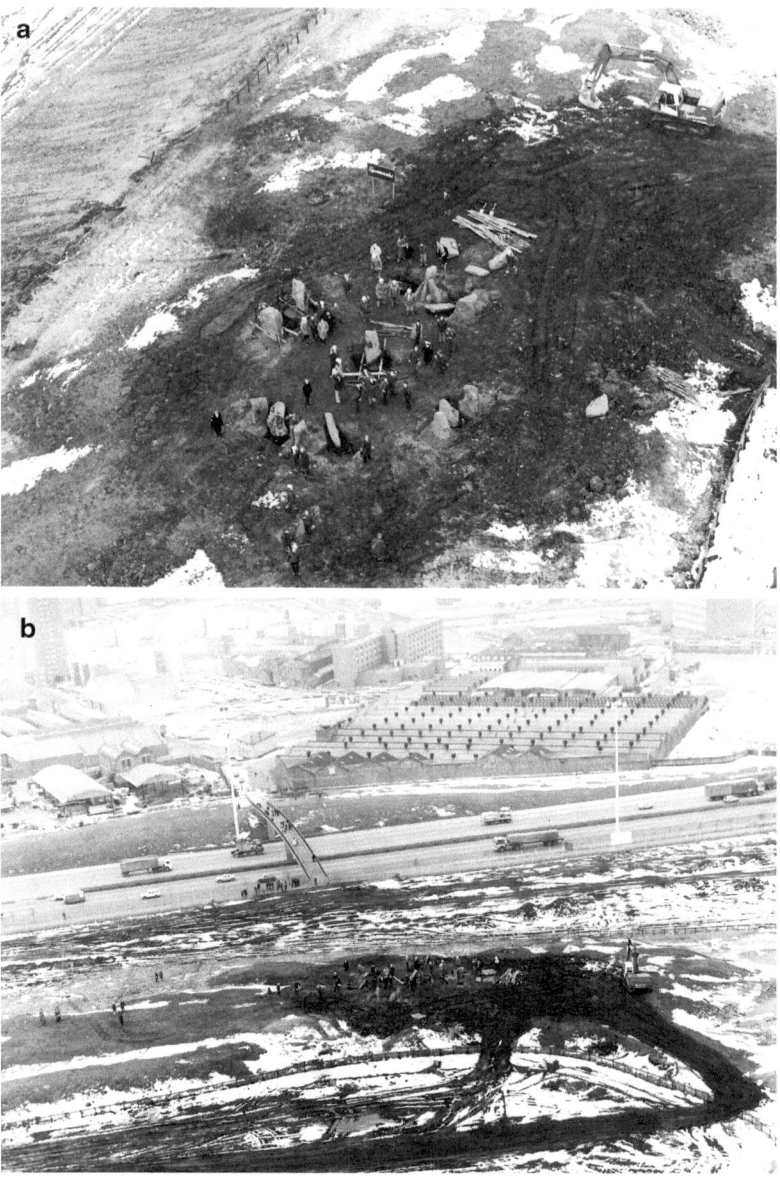

FIG. 7.24 (a) The circle completed. (b) The helicopter pulls out. Gavin Roberts's wife and daughter are on the footbridge; young Sarah was convinced the helicopter was a train, partly because it looked like one, nose-on, but also, it was going back and forth carrying things. Royal Navy photograph

Fig. 7.25 Left to right, Dr. Euan MacKie, Prof. Alexander Thom and the late Bill Braithwaite. Note the displays of *Pioneer* Jupiter and *Viking* Mars photographs, on loan to the project from ASTRA for use in schools

operation about to start, sir, please take my seat on the aircraft," and as it flew off, turned to his driver and said, "Take me to the party." What the MoD inspector thought, when he found he was to participate in building a megalithic circle, is not on record.

We had of course made a point of inviting the ground crew, little realizing how unusual that was. Later we were told of a more typical incident, where a retired admiral in the village of Dunure had asked for a detachment of sailors to move his yacht's mooring in the harbor. After hours chest-deep in mud at low tide, in the middle of winter, they had been left soaking on the harbor wall while the ex-admiral took the officers for a drink; and this time, too, they were expected to wait outside the building, with snow threatening again, for transport that wasn't coming for more than 3 h.

John Braithwaite reported this to Keith Fraser, who authorized the Special Projects secretary, Pia Pisaneschi, to issue John with a float from petty cash to take the ground crew to the George Hotel nearby. The Operations Manager, Alan Montgomery, was ordered to make sure everyone else returned to their desks, and the Parks Department management collected up the remaining bottles of wine and left.

At the level of adrenaline we had all been operating under, there was no chance whatever of that. Looking around, Alan saw the entire staff of Special Projects putting on their coats. "Where do you think you're going?" he asked, and we told him. I was last out, and from the top of the stairs, Alan called after us, "You do realize this may result in disciplinary action..."

"Get your coat, Alan," I shouted back, "and close the building!"

Meanwhile, in the George Hotel, John was coping with the disbelief of the ground crew. "But you had dinner with the C.O. last week!" We had noticed that the officers and ratings had kept far apart at the reception, but still didn't appreciate the total class separation of ranks in H. M. forces at that time. "Listen," said John, "in a moment that door will open and in will come Duncan Lunan, Gavin Roberts and the rest of the happy multitude..." And right on cue we did.

From there and a succession of other bars I went off to do more and more interviews with the press and media. My wife came with me to the last one, which was with Radio Clyde in the early evening, and went so well that afterwards I had a monthly radio spot there for the next 18 months. As we were heading back to rejoin the gang in the pub, Linda said to me, "That was your last interview today, wasn't it?" When I confirmed that, she said, "In that case, I have something to tell you. I haven't told you till now, because you'd have been so big-headed with it you'd have been insufferable.

> You know when Archie Thom went up the hill to join you? I was left with Professor Thom in the Range Rover, surrounded by children cheering like mad every time the helicopter came over with another stone. We didn't talk much, but when Dr. Thom came sliding back down in the kilt, in the snow and the mud, the Professor said, 'Can you imagine Glyn Daniel doing that?' And when Archie Thom got back into the car, the first thing the Professor said to him was, 'Don't you wish Glyn Daniel was here to see this?'

He may not have been proud of the circle, but I was proud enough for both of us.

Next day the blizzards returned, and operations were confined to the recovery of the nets. The delay gave time for the completion of the time capsule, which contained also the press coverage of the event itself. My idea had been that if some latter-day Gerald Hawkins ran the alignments through his computer, when the origin of the circle was forgotten, the two dates for Rigel would alert him. If he probed the ground, the concrete foundations would tell him which date was correct; and if he noticed that the foundation of the central stone was elongated, scanning it would reveal the time capsule, where the books would tell him what we knew about the ancient sites and their builders. But there was no budget to fill the capsule with nitrogen, or for calcium chloride crystals to absorb moisture, nor were the books and newspapers special editions on acid-free paper. They were simply sealed in plastic bags and the capsule was sealed shut by its maker, David Sneddon of the S.T.E.P. Biogas project; it's unlikely that anything survives by now.

On the 22nd Gavin supervised the erection of the helicopter stones, joined for the last few by John Braithwaite and myself; the capsule was ceremonially buried, the working photographs were signed by the Sportsworks and Astronomy Project teams, and at its first stage of construction, the circle was complete. The late Ian Downie had now joined the project, and in the following weeks he and the official photographer, John Gilmour, completed major photographic studies of the circle in its freshly completed form (Figs. 7.26, 7.27, and 7.28).

The story made the first page of *The Scotsman*, and was covered with photographs by the *Glasgow Herald*, the *Daily Record* and others, also nationally without pictures by the *Daily Mail* and *Daily Telegraph*. We didn't realize that the publicity was attracting unwelcome attention in the new Conservative government, until our denunciation in the House of Commons. Until then, with the project fully manned with genuinely unemployed, including two registered disabled and one a single mother, we thought we were fulfilling the objectives pretty well, and in addition to the success of the helicopter operation, our work in schools was now in great demand and that alone justified our existence. Or so we thought.

FIG. 7.26 The circle newly built, looking west to Glasgow University (*left of center*) and the Summerhill (Photo by Ian Downie)

I was so stressed during the operation that it wasn't until I was looking at the press coverage that I realized which helicopter we had been sent. I had seen 'Zero One' before. I had once been picking up David Proffitt from the camp gate when 01 took off, only to hover for a long time over the apron. "Why isn't he flying off?" David wondered, and I jokingly said, "Perhaps he's scared to." "With Zero One, you could be right," said David, going on to explain that in Naval Air Squadrons the designation was given to the oldest aircraft, the 'hangar queen' normally used as a source of spares. Once when he was on aircraft carrier duty, a Zero One fighter-bomber had been launched only to get into immediate difficulties. The crew ejected, but the machine fell into a menacing circuit of the carrier, while the gun crews were scrambled to their posts to shoot it down.

I don't know whether anyone in the MoD or the Navy had thought we might write the helicopter off for them. However, David later reported, HMS Gannet's Zero One had apparently benefited from the powers of stone circles, gaining a new lease of

Fig. 7.27 (a) The newly built circle on the skyline from 'Dead Man's Gulch' (Photo by John Gilmour). (b) The newly built circle looking east towards Townhead (Photo by John Gilmour). (c) Newly built circle looking southeast. Townhead including Townhead Church on *left*, Glasgow Royal Infirmary and Cathedral to *right*, with Cathkin Braes to the rear (Photo by John Gilmour)

Fɪɢ. 7.28 (a) Author on the motorway bridge over the M8, where the former Forth and Clyde Canal spur was; derelict railway sidings behind. Note the old 1960s jacket and the No. 1 badge from *The Prisoner*, which was being repeated at the time. (b) On the bridge, scene of a tragic incident in 2002 (Chap. 10). Notice the flared jeans, de rigeur in the late 1970s but quite impractical for quarries and muddy hillsides (Photos by John Gilmour)

life. "It's turned into not a bad bomber." Sadly, though, the effect was temporary. A year or so later, she developed engine trouble and put down on the beach on the island of Arran, where she was flooded by the tide before recovery could be mounted and was deemed uneconomic to repair.

Over the span of the project, we came to believe that the Sighthill stones were the first astronomically aligned stone circle to be erected in the British Isles for at least 3,000 years. At first I made the claim to see if it would be challenged, but for more than 30 years now it hasn't been. Quite recently, I've learned that in the early nineteenth century Lord Clermont built one at Ravensdale, County Louth, in Ireland – on land then belonging to members of my mother's family, remarkably enough. It's an ellipse with the long axis pointing north–south, four stones roughly aligned with the cardinal points and four perhaps for the solstices, though as it's in woodland, on a slope, and the stones are symmetrically placed, I doubt if it's astronomically accurate.

Late in the nineteenth century the Astronomer Royal, Sir Norman Lockyer, was convinced of the astronomical alignments of ancient sites, and gave advice on the building of stone circles

FIG. 7.29 (**a**, **b**) Bardic circle at Plas Newydd, Llangollen (Photos by Linda Lunan, 2011)

around Llangollen in Wales for the annual Eisteddfods. Tony Crerar assured me that they weren't accurate, and from what I can gather they were impermanent, moved from one location to another year by year (Fig. 7.29). As mentioned earlier, outside the Spaceguard

FIG. 7.30 (a) Part of the limestone circle at the Spaceguard Observatory, Knighton, Powys, Wales. (b) Taurus stone at Spaceguard Observatory (Photos by Linda Lunan, Sep 2011). (c) Princess Diana stone at Spaceguard Observatory

Observatory in Powys there is a symbolic stone circle of limestone blocks, removed from Ely Cathedral during renovations, and dedicated to Princess Diana (Fig. 7.30c). That inscription is thought to be more recent than the circle, but even so, it's all more recent than Sighthill, and although several of the stones have carvings of planets and constellations, they're not aligned accurately to rising and setting points (Fig. 7.30b). Quite a number of stone circles have been built in recent years, including replicas of Stonehenge on latitudes where they won't work; but as far as I know, we are the first to have done what we did. Some significant things have been learned as a result, and they'll be explained in Chap. 8.

References

1. Jordan, S., Patterson, W.: Jeff Hawke, Overland, H3979, Daily Express, 17/12/66–24/5/67. Reprinted in: Jeff Hawke's Cosmos, 6(3), 42–76 (April 2011)
2. Thom, A.S., Ker, J.M.D., Burrows, T.R.: The Bush Barrow gold lozenge: is it a solar and lunar calendar for Stonehenge? Antiquity, 62(236), 492–502 (Sep 1988); Smith, G.: Relic ruined by experts is unique Druid calendar. Daily Telegraph, 1 Sep 1988. Actually Dr. Thom and his co-authors wouldn't have mentioned the Druids, for whom Stonehenge was more ancient than they are to us
3. MacKie, E.W.: The Prehistoric Solar Calendar: an Out-of-Fashion Idea Revisited with New Evidence, op cit
4. Anonymous: Herald Diary: The multi-storey Sighthenge. Glasgow Herald, 20 July 1978
5. McKay, T.: The magic circle hits a spell of trouble. Daily Record, 13 Nov 1978
6. Holder, G.: The Guide to Mysterious Glasgow, op cit

8 Events on Site

We know by the Moon that we are not too soon,
We know by the sky that we are not too high,
We know by the stars that we are not too far,
And we know by the ground, that we are within sound...
> – "Gower Wassail" (traditional) [1].

In my Radio Clyde interview, Dave Jamieson had asked when the first visible event at the circle would be. Midsummer solstice, was the reply – June 21. But if the stones were set in concrete, what if the midsummer Sun didn't rise in line with them? "Well," I replied, "I don't know about the rest of them, but I'm for South America. I might be able to flannel it with the Parks Department, but not with Professor Thom."

To use the circle, the observer is supposed to stand on the far side from the event, so that the center stone occults the marker stone and he or she sees the heavenly body rise or set above it. The shadow of the marker stone would fall on the center stone, and the shadow of the central stone towards the observer, demonstrating the event even to a large number of onlookers. The ground falls away on the south side of the circle, and to raise the observers to the correct positions there, the intention was to create a low bank, so incorporating in a sense at least one feature of Stonehenge. In completion of the park the center of the circle, the perimeter and the four segments leading to the stones were to be paved with "cossie sets" of granite (Fig. 8.1), between four arcs of grass, corresponding in a way to the beds of stone found by Mackie in his excavations at Kintraw and Cultoon, also found in the small stone circle at Temple Wood (Fig. 8.2).

The crucial test (weather permitting) would be the midsummer solstice. The geometry of the situation is such that if even one solar alignment was correct all the solar and lunar ones should be correct (but see below!), and only the Rigel alignments would

D. Lunan, *The Stones and the Stars: Building Scotland's Newest Megalith*, Astronomers' Universe, DOI 10.1007/978-1-4614-5354-3_8, © Springer Science+Business Media New York 2013

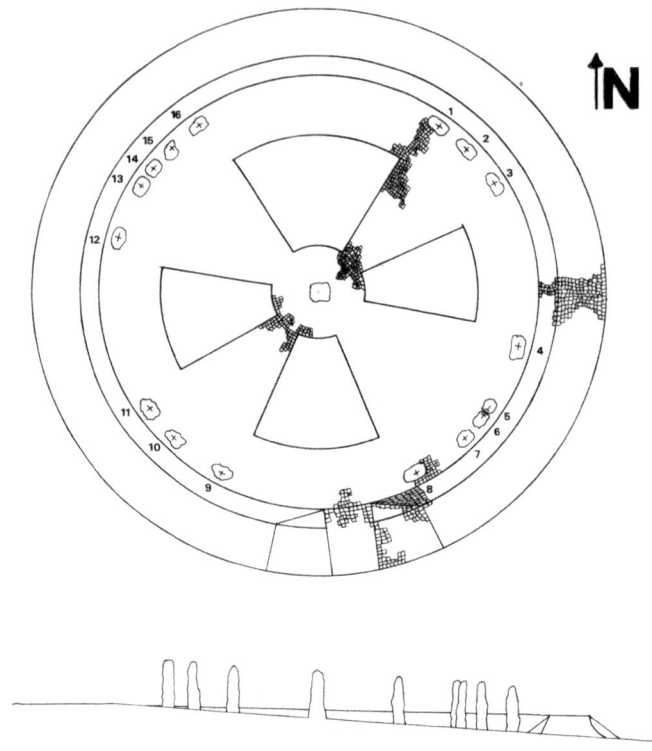

FIG. 8.1 Planned landscaping of the stone circle; for the key to the stones see Fig. 6.23 (Drawing by Richard Robertson)

remain to be verified. A sizeable company of 30 people gathered on the hilltop on the early morning of Thursday, June 21, but the sky was completely overcast and after 10 min it began to rain, and Thursday and Friday evenings were no better. Around the solstice, the Sun's declinations change little over a few days, but Friday and Saturday mornings were so atrocious that no one made the attempt to be present. On Saturday evening, when the attendance had dropped to just myself, the sunset was clear enough to confirm that the alignment of the circle was approximately accurate (Fig. 8.3). Although by that time the sunset point was moving south, and there were too many clouds for precise observations, the setting position was still well within the accuracy we had been able to achieve without distant foresights. (The only possible

FIG. 8.2 Small stone circle at Temple Wood, Argyllshire, with central burial kist and bed of stones (Photo by Chris O'Kane)

FIG. 8.3 Sunset, Saturday June 23, 1979 (Photo by author)

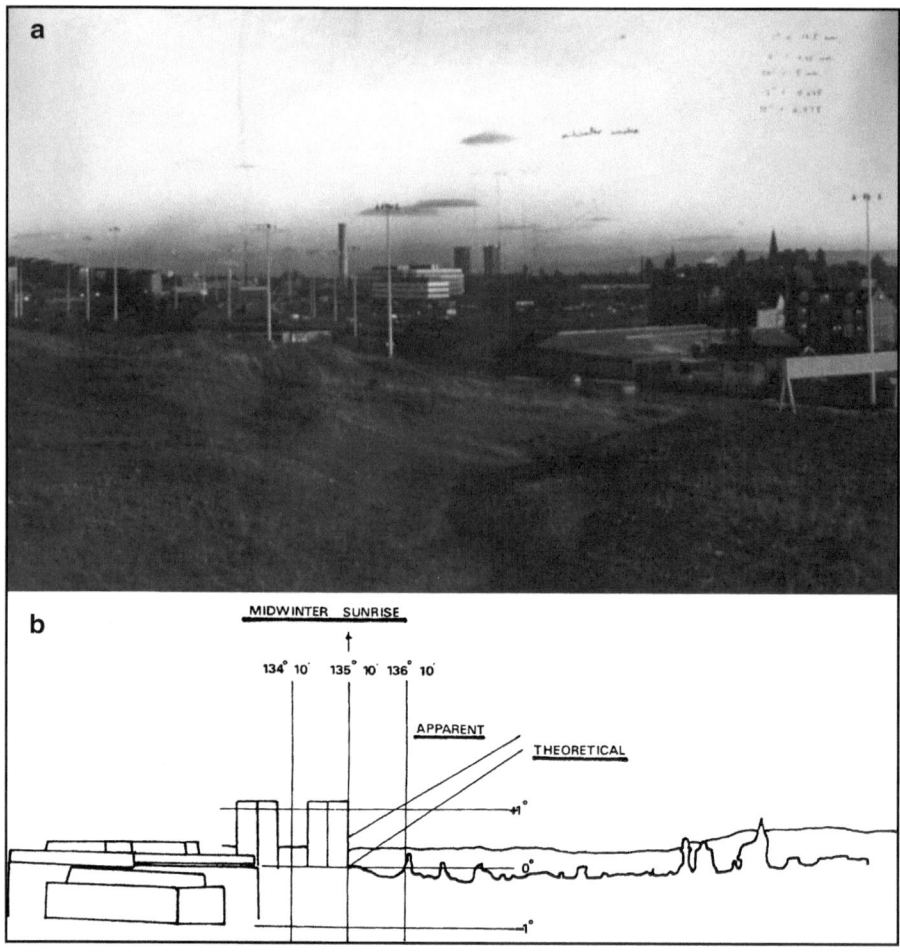

FIG. **8.4** (**a**, **b**) Midwinter sunrise prediction. Working photograph by Gavin Roberts, drawing by author (Redrawn by Richard Robertson). (**c**, **d**, **e**) Midwinter sunrise, 1979 (Photos by Gavin Roberts)

one, Ben Lomond, was just too far north to be usable – Fig. 8.5c). I mailed copies of Fig. 8.3 to the Thoms, MacKie and Roy, saying 'the plane tickets to Buenos Aires have been cancelled.'

Midwinter sunrise, 1979, was observed and photographed by Gavin Roberts, and proved to be almost exactly as predicted (Fig. 8.4). It turned out later that there was a considerable element of luck in that, though not because of any error in the calculations. The haze on that part of the skyline, virtually a permanent feature that had given a great deal of trouble in the survey, was mostly

FIG. 8.4 (Continued)

industrial and due in large part to steelworks in the Clyde Valley. Cutbacks the following year brought Tinto Hill into view, alarmingly close to the midwinter sunrise alignment but hidden from us throughout the project.

The weather had closed in by sunset, and no further observations were possible that year. With the termination of all STEP schemes the Astronomy Project was wound up in February 1980, and practical difficulties prevented any observations in midsummer; in midwinter the weather was again unsuitable. Project members had participated in the restoration and reopening of Airdrie Public Observatory, and the contemporary rising of Rigel was studied during the winter by Paul Benson, one of the curators, but he concluded that the thickness of the haze and interference of city lights made Rigel impossible to observe below 20–30° altitude.

Midsummer sunset of 1981 took place on a spectacularly clear evening and was most informative. Instead of setting on the pitched roof of the White Horse distillery as predicted (Fig. 8.5a), the setting Sun passed behind the peak at the end of the roof only partly hidden and re-emerged fully to set on the natural skyline beyond (Fig. 8.5b, c, d). This was virtually the only event not located on a natural skyline by calculation, and as it was still over the central stone, and even over the marker stone (Fig. 8.6a), with its shadow falling on the central stone as intended (Fig. 8.7), it could be said that the discrepancy improved the functioning of the circle, though the calculated position was apparently too far to the left.

The following morning was extremely tantalizing. There was a low bank of cloud along the northeastern horizon, and the Sun seemed determined not to leave it. When it did eventually reach the top of the cloud layer, there was insufficient contrast in the photographs for its position to be fixed with precision. By then it was so far from its point of rising that to be accurate both the 'theoretical' and 'apparent' tracks in the working photograph would have had to be shown as curves. All that could be said was that to be photographed over the central stone in the position shown, the Sun must have risen fairly close to the predicted point.

Weather again prevented observations in midwinter 1981, and practical difficulties again intervened in the summer of 1982. These frustrations were to have a wry counterpart years later, when Archie Roy was asked to design basic layouts for two circles

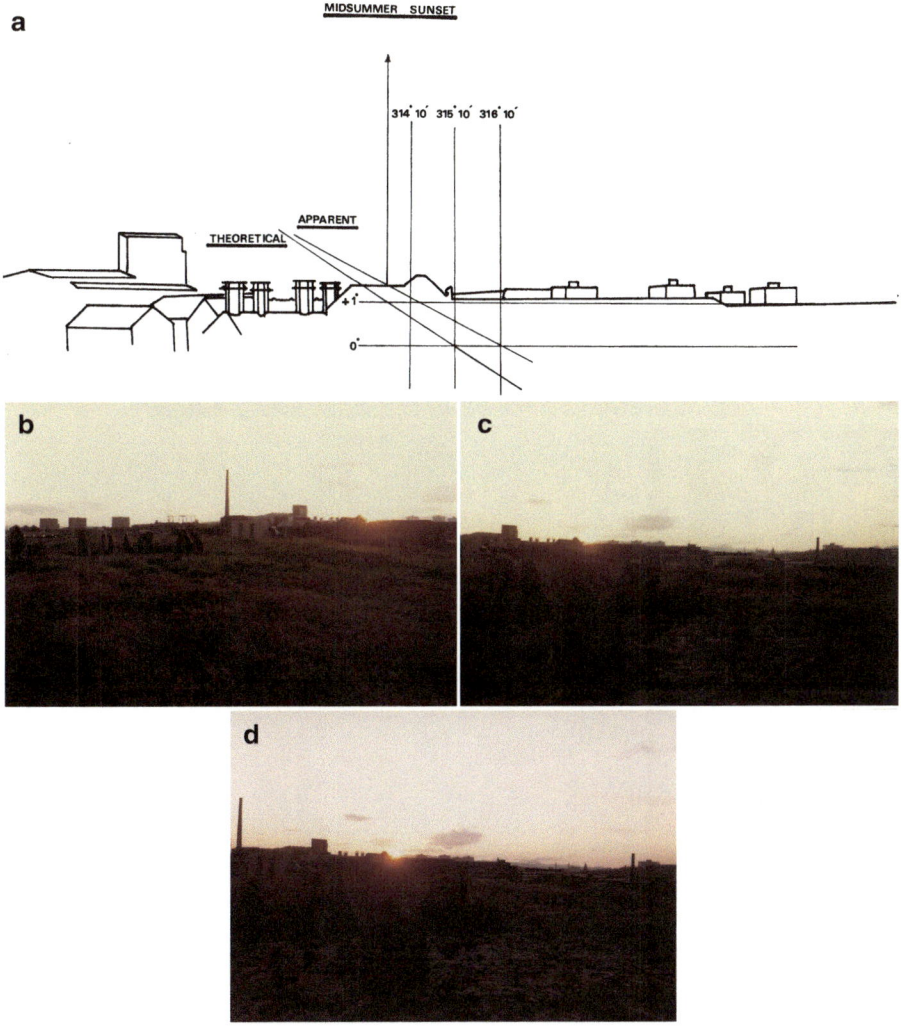

FIG. 8.5 (a) Midsummer sunset prediction. Drawing by Richard Robertson, based on the working photograph. (b, c) Midsummer sunset, 1981. (d) Fig. 8.5c enhanced. Note Ben Lomond to right of center (Photos by author)

of concrete blocks, one at Linwood on the far side of Paisley and the second at another Sighthill, on the east side of Edinburgh. He wasn't involved as deeply in the constructions as we had been in ours, but he did go out to Linwood at midsummer sunrise in hopes to see it. However, he couldn't gain access because the local council had sown the site with expensive grass seed and surrounded it with a security fence.

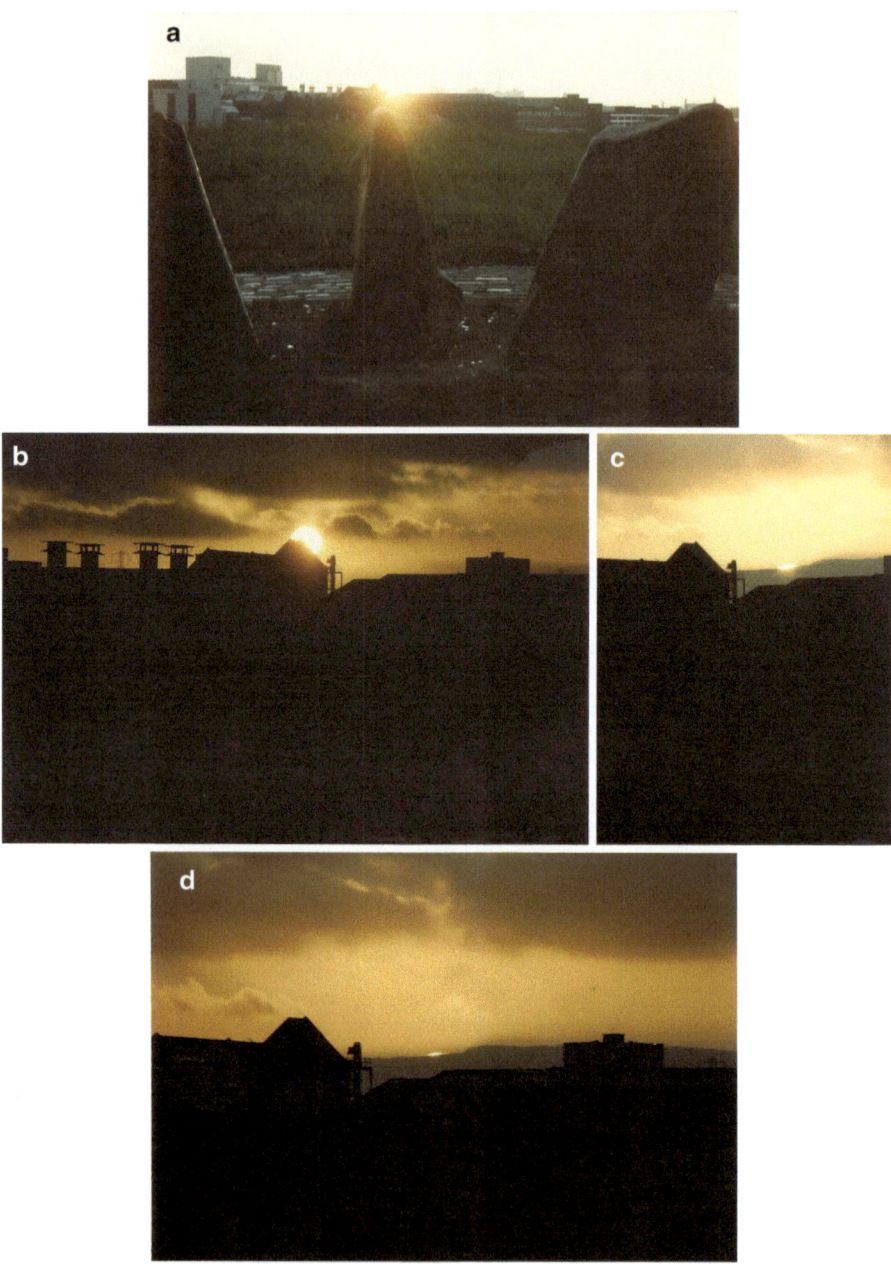

FIG. **8.6** (**a**) Midsummer sunset, 1986, over the marker stone (just) (Photo by author). (**b**, **c**, **d**) Midsummer sunset 1986 (Photos by Chris O'Kane)

Fɪɢ. **8.7** Author with midsummer sunset shadow, 2010 (Photo by Graham Gardner)

December 1982, however, at last allowed a check on the elusive midwinter sunset alignment, which was predicted to be where St. Andrew House on West Nile Street cut the horizon of Gleniffer Braes, beyond Paisley (Fig. 8.8). There were clouds on the skyline in the southwest, but they weren't a problem because the descending Sun's track proved to be half a degree higher than predicted (Fig. 8.9). From the glow in the clouds, indeed, it might have reappeared briefly to the right of the tower which cuts the skyline (Fig. 8.9d), in which case the precise alignment would (fortuitously) be given by the right-hand edges rather than the centers of the marker and central stone. However the shadow pattern was as predicted, with the shadow of the marker falling on the central stone (Fig. 8.10), and the shadow of the central stone falling on the midsummer sunrise stone almost opposite (Fig. 8.11).

Despite the good weather in June 1983 it again proved impossible to have an observer present on the morning of the solstice. Robert Law of the Coats Observatory in Paisley (and now of the

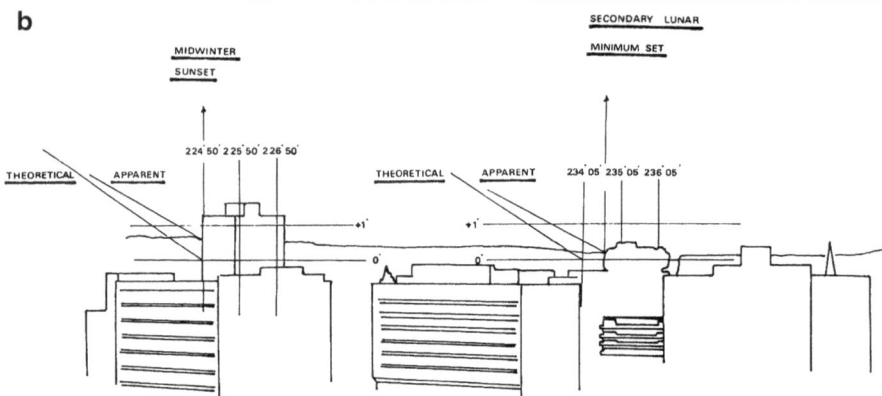

FIG. 8.8 (a, b) Midwinter sunset prediction (Working photo by Gavin Roberts, drawing by author) (Redrawn by Richard Robertson)

Mills Observatory, Dundee) photographed the sunrise on the morning of June 19, obtaining some spectacular shots of the Sun over the central stone [2]. But not knowing what was needed, he didn't fix the rising point on the horizon.

In an article submitted to the U.S. magazine *Tournaments Illuminated* in 1986, I wrote, "Believe it or not, due to weather and other problems, after 7 years we still don't have a precise record of midsummer sunrise," [3] though it was the main feature

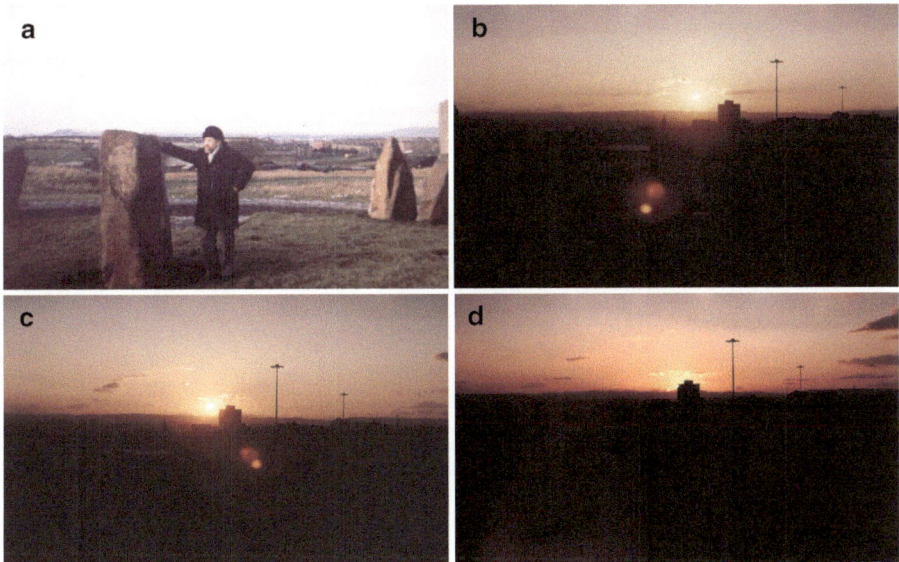

Fig. 8.9 (**a**) Northwest view with the late Prof. Oscar Schwiglhofer, founder of ASTRA, midwinter sunset 1982 (Photos 8.9–8.14 by author). (**b, c**) Midwinter sunset, 1982. (**d**) Midwinter sunset aftermath, 1982

Fig. 8.10 Midwinter marker shadow on central stone, 1982

FIG. **8.11** Central stone shadow on midsummer marker, midwinter 1982

of the design! In fact it was to be 13 years before I saw it. By 1989, a large bush had grown up on precisely the line where the rising sun should be (Fig. 8.12). After I mentioned that in a talk to the Airdrie Branch of ASTRA, the late Danny Kane (who was in the paint business) secretly addressed the matter with a pot of creosote, and by 1992, the bush had died back sufficiently to let the sunrise be seen from the viewpoint on the other side of the circle (Fig. 8.13). Once again, though the shadow pattern fell on the central stone and the midwinter marker as intended (Fig. 8.14), the rising point was well to the right of prediction (Fig. 8.15).

By this time the discerning reader will have realized what was happening, as indeed I had. Midwinter sunrise track was almost exactly as predicted (when first seen), the midwinter sunset one was about half a decree higher, and the midsummer sunset one about two-thirds of a degree higher. The most obvious factor of difference is air temperature, which affects the apparent upward displacement of the Sun's image by increasing the effect of atmospheric refraction. With the helicopter operation pending, pressure of time forced me to use the same average values for atmospheric refraction, midsummer and midwinter, in all four

FIG. **8.12** Midsummer sunrise concealed, 1989

FIG. **8.13** Midsummer sunrise revealed, 1992

computations – obtained from Thom's books. I didn't allow for the variations in refraction, from sunrise to sunset, in the relatively dense air over the city. At sunrise the air is cooler than average, and the Sun's track appears lower, moving its apparent position to the right; at sunset, when the air is warmer than average, the

FIG. **8.14** John Stark of ASTRA witnesses central stone shadow on the midwinter sunset marker, midsummer solstice, 1992. Glasgow University spire left of center

FIG. **8.15** Midsummer sunrise, 1992, overlaid with predicted alignment (Photo by author, drawing from working photograph by Richard Robertson)

FIG. 8.16 (a, b) Midsummer sunset, 1992, compared with prediction (Photos by author, overlay by Richard Robertson)

track is higher and again the Sun's disc is displaced to the right. The variation from theoretical values will be still more enhanced by the relative pollution, and therefore higher than average density, of the air over the city.

Midsummer sunset that night provided another illustration (Fig. 8.16), and the explanation was confirmed by observations of midwinter sunrise over several years, where the rising point of the

Sun moved several diameters left and right with the differences in air temperature.

Archie Thom pointed out that the Sighthill megalith is particularly vulnerable to such effects because its skyline is in a sense too good. The effects of refraction are not only much greater but also much more susceptible to variation between 0° and 1° altitude. The Neolithic builders must have been aware of the effect, because they typically used sight-lines with elevations of 2–4°, where variations in refraction were much less, and if they used a true horizon it tended to be a sea horizon, with a cooling, calming effect, and ideally looking down from an elevation of 1° or 2°, as at the Scalisaig stone on Colonsay (Fig. 4.23).

If the observations had agreed exactly with prediction, relatively little would have been learned. The original idea was that if the features of the Sighthill circle were all based upon surviving Neolithic sites, and yet it functioned as an observatory and the project members could testify that it had been intended from the outset to be one, then anyone maintaining that the ancient sites were not observatories would have to point to some significant differences. Now, however, it was obvious that if the Sighthill alignments had been determined by observation, rather than calculation, the sunset alignments would have been a great deal more accurate than they are, even within the limitations of a circle 40 ft across. If horizon features were used as 'foresights,' as Thom suggested, then much greater accuracy would indeed have been attainable. For example, I had calculated the most northerly setting of the Moon (at lunar standstill) to be where the natural line of the hillside met the stepped roof of the right-hand distillery building (Fig. 8.17). In Neolithic times, no doubt the priority would have been to find a site further north, from which the significant moonset would be seen in the notch formed by Ben Lomond's rim on the skyline – a shift of less than 2° in azimuth is needed. Then, observing the rim of the Moon rather than its center, accuracy to within a minute or two of arc, could indeed have been achieved as Thom maintained.

The 'errors' actually enable us to make an important point, which otherwise wouldn't be obvious. As explained above, at the ancient sites nothing happens exactly where it did, because the tilt of Earth's axis has changed by half a degree meantime. It's enough to let the skeptics say that the calculated alignments are

FIG. 8.17 Major lunar standstill, northerly set, prediction, above the White Horse Distillery sign (Working photograph by Gavin Roberts, drawing by author)

spurious because the claimed accuracy can't be achieved by observation alone. Ah, but it can. Sighthill would demonstrably have been much more accurate if it had been aligned by eye.

The lunar events were eventually to make the point still more strongly. As explained above, since the Moon's orbit is inclined at about 5° to the ecliptic, which is the plane of Earth's orbit projected onto the sky, and since the Moon's orbital plane precesses around the sky with a period of 18.61 years, during that period there are two 'standstills' when moonrise and moonset reach their furthest north and south for 9.3 years. We had just missed a minor standstill when the Astronomy Project was set up, so there were to be no lunar events at Sighthill until September–October 1987. Then we would get four events in a single month, with the Moon rising and setting at its furthest south in September, then furthest north 2 weeks later, at the major standstill.

In the 1986 article, I wrote, "Weather, be kind to us then!" But it was not to be. Reckoning that to the accuracy of my circle, 6 months either side of the event would still be close enough, I went up there, alone or with friends, on each occasion of the 48 when

FIG. 8.18 First attempt at lunar major standstill northerly rise. The Moon becoming visible over Sighthill Cemetery, too long after the event to plot the rising point (Photo by Mark Runnacles, November 5, 2005)

it seemed the weather might allow a glimpse of an event, and not a one did we see. At the minor standstill in 1997, I was working as a Precognitions Agent, taking defense witness statements for Angela Mullane, solicitor wife of my old friend Chris Boyce. It was valuable experience for a writer – "You meet all kinds," as they say – but it made it impossible to plan my life around moonrises and sets. When Tony Crerar came up from Wales, hoping to whisk me off to the Highlands to help verify predictions of Thom's, the most he actually could do was drive me to the West End in search of a recalcitrant witness.

For 2006–2007, I planned a campaign with Mark Runnacles of the *Daily Record*, extending our 'window' to a year on either side of the major standstill. After several setbacks with clouds (Fig. 8.18), we managed to photograph the Moon setting at furthest north, and at furthest south (Figs. 8.19 and 8.20). I saw the Full Moon rise at furthest south, precisely on the predicted alignment (Fig. 8.21) with perfect clarity on a night when Mark couldn't be there; my attempt to photograph it with a pocket camera failed,

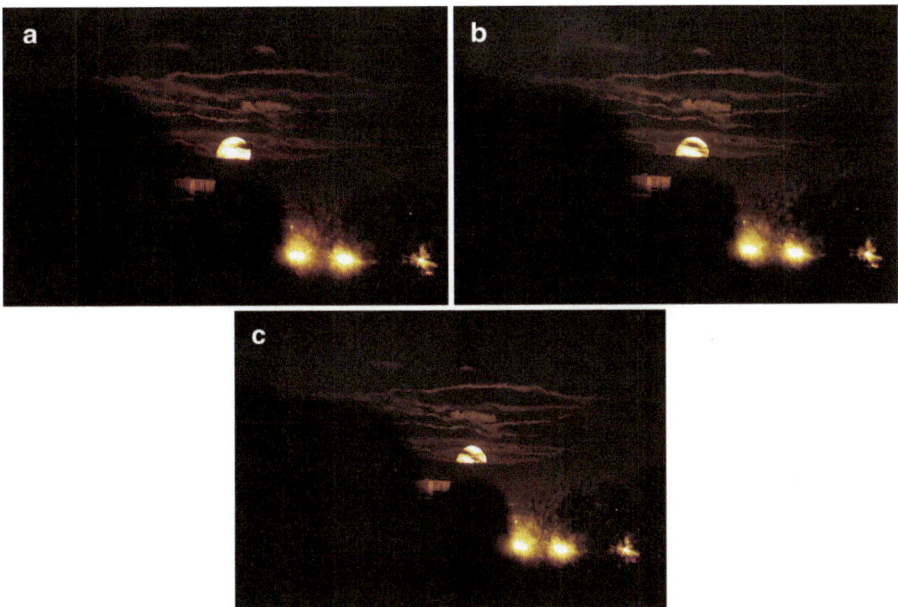

Fig. 8.19 (**a**, **b**, **c**) Moonset at furthest north, over the silhouette of the marker stone (Photos by Mark Runnacles, February 9, 2006)

and all I have for that alignment are snapshots taken by my colleague Bob Graham in a different month, well after moonrise when it finally cleared the clouds (Fig. 8.22).

Within the limits of accuracy of the stones, the southerly events and the northerly moonset appeared to be where they should be (Figs. 8.19, 8.20, 8.21 and 8.22). But northerly moonrise was to be a shock – not visible until it emerged from behind the tower blocks, on a much lower track than predicted (Fig. 8.23). Projected back, it met the true horizon at the theoretical rising point (Fig. 8.24), but in real life, when it rose above the true horizon, it would have been far to the right of the prediction, almost out of line with the stone!

Once again the source of the error became clear, once I knew it was wrong. Under pressure to finalize the calculations, I had assumed that the rising and setting paths would all make roughly the same angle with the horizon. I knew that wouldn't be the case at higher latitudes, but I didn't think it would make a big difference at the latitude of Glasgow. Instead of doing it graphically, I should have calculated the expected positions of the Moon (and

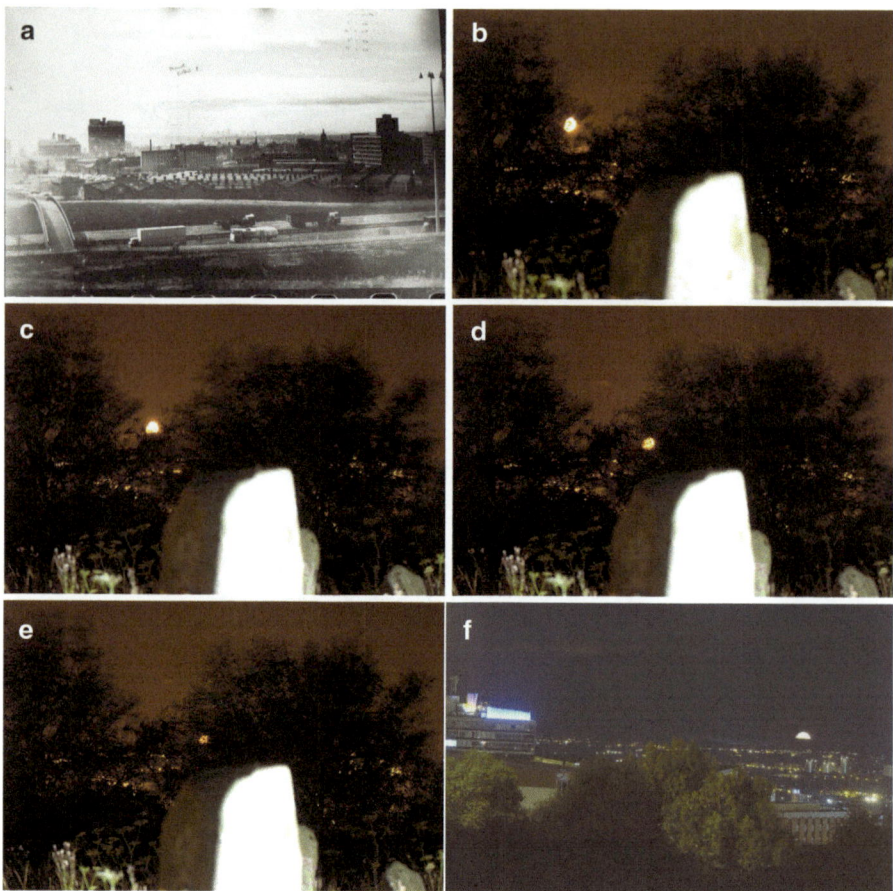

FIG. 8.20 (a) Extreme southerly moonset prediction, on the horizon of Gleniffer Braes, between the College of Building and Printing (*left*) and St. Andrew House (*right*) (Working photo by Gavin Roberts, drawing by author). (b, c, d, e) Moonset at furthest south, now concealed at the horizon by trees. (f) Comparison image of the Moon on the clear horizon of Gleniffer Braes, from several diameters west of the circle. College of Building and Printing to left (Photos by Mark Runnacles, August 23, 2007)

for greater accuracy, the Sun) at several places along the track, though to be fair to myself, none of the experts to whom I showed the graphs on the working photographs noticed anything wrong.

As well as the solar and lunar events, the Sighthill circle marks the rising of Rigel, at the foot of Orion (Fig. 8.25), as it is now and as it was at the end of the megalith era in 1800 B.C. (Fig. 8.26) – see Chap. 5. The present-day alignment fell on the end of the historic Townhead Church (Fig. 8.27), which was controversially

FIG. 8.21 Extreme southerly moonrise prediction, on the horizon of Cathkin Braes, over Glasgow Royal Infirmary to the right of the Cathedral spire (Working photograph by Gavin Roberts, drawing by author)

FIG. 8.22 (a, b) After major standstill southerly moonrise. Moon curving right at low altitude. For the attachment to the central stone, see Chap. 10. The airborne blobs in Fig. 8.22b are pollen being released as night fell (Photos by Bob Graham of North Lanarkshire Astronomy Project, July 27, 2007)

demolished in the 1990s, though the steeple was restored and preserved (Fig. 8.28). The poet Brian Finch wrote, in a series of satirical haiku mocking Glasgow's pretensions as 'City of Architecture and Design': [4]

FIG. 8.23 (a, b) Extreme northerly moonrise, major standstill (Photos by Mark Runnacles, September 14, 2006)

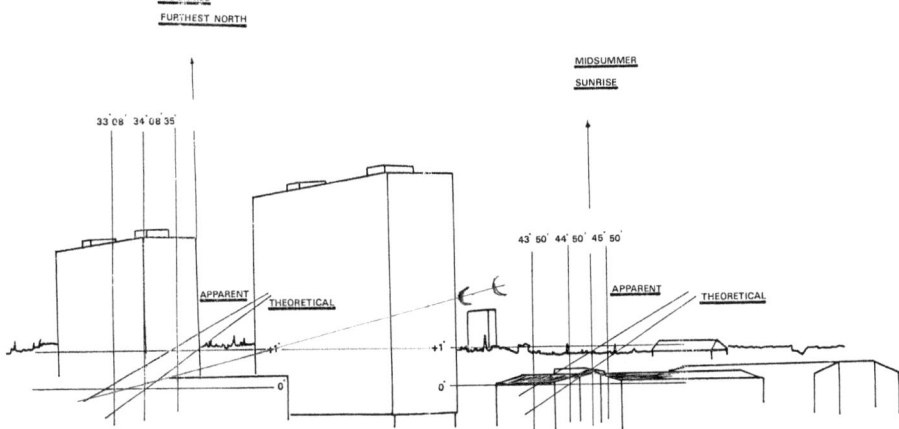

FIG. 8.24 Moonrise furthest north comparison (Original drawing by Richard Robertson from working photograph, sketch by Duncan Lunan)

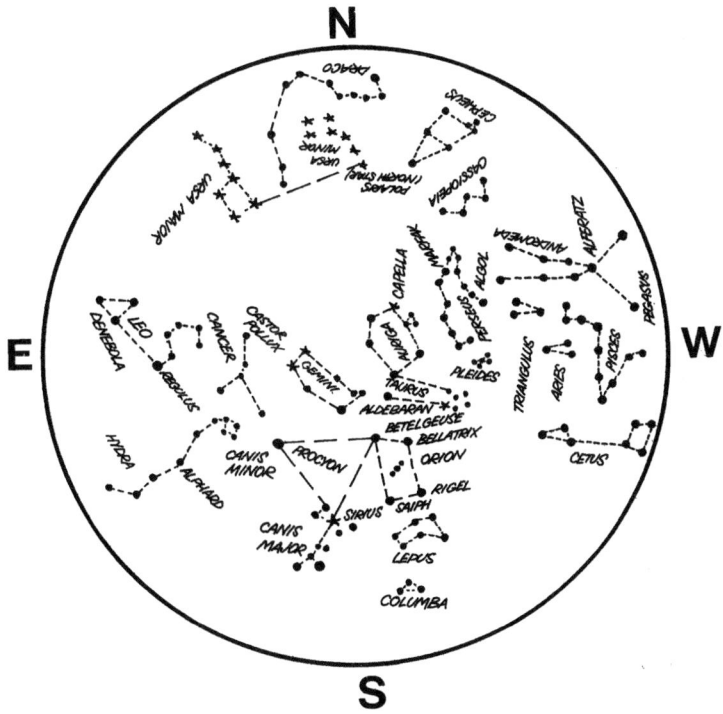

FIG. 8.25 The night sky about 9 p.m., mid-February, drawn for the British Isles by Jim Barker, 1983, for author's monthly column 'The Sky Above You.'

FIG. 8.26 (a) Rigel rising in 1800 B.C., to the left of where the incinerator chimney is now. (b) Rigel rising, late 1970s, on the end of the Townhead Church (Working photographs by Gavin Roberts, drawings by author)

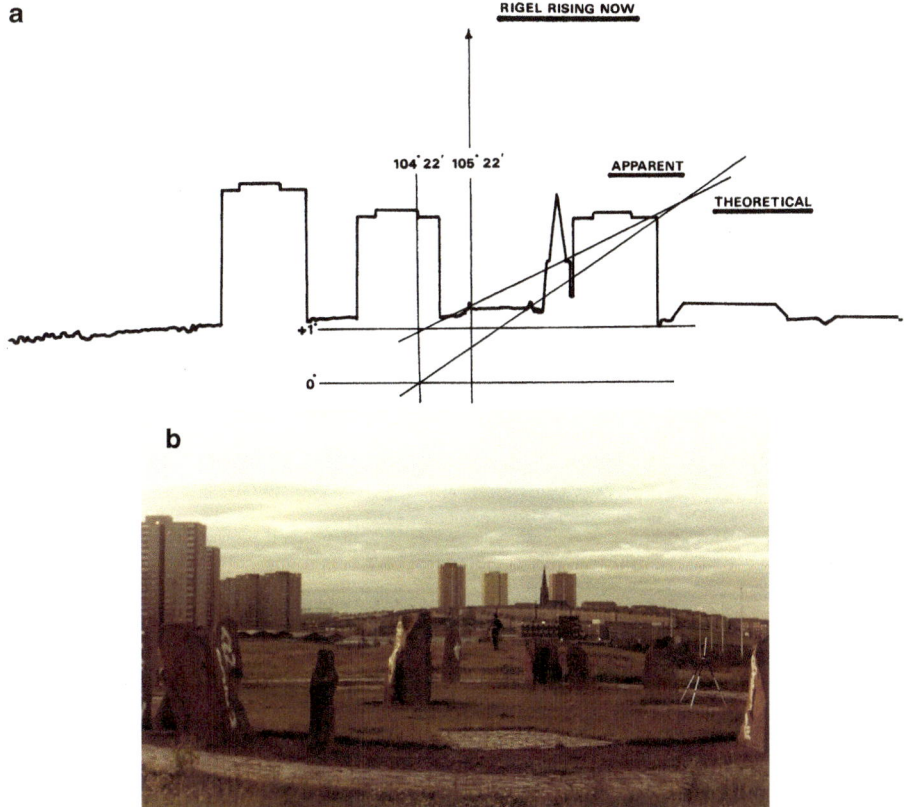

FIG. 8.27 (a) Rigel rising prediction. Drawing by Richard Robertson, from Fig. 8.26b working photograph. (b) Looking east towards Townhead Church before midsummer sunrise, 1989. Author in middle distance (Photo by Ian Downie)

> Tounheid Kirk's awaa
> The toun council caa'd it doun
> Still the steeple stauns.

After Paul Benson tried to photograph the star at rising without success, in 1986 I wrote, "Because of the lights and haze of the city, no-one has seen Rigel rise over the marker; and since the rising points of the stars change fairly rapidly, due to the wobble in the Earth's axis known as Precession of the Equinoxes, it may be that no-one ever will [3]."

That proved unduly pessimistic. As with the clearing of the industrial pollution over the Clyde Valley in the early 1980s, in

Fig. 8.28 Restoration of Townhead Church spire, 2001. The stone circle is on the green hilltop at upper left of center

the twenty-first century Greater Glasgow has adopted a policy towards light pollution that will see all municipal lighting made dark-sky compatible, on a replacement basis, within 20 years. Already the difference is marked, and whereas in Mark Runnacles's 2006 photographs the sky background is orange (Fig. 8.23), when he photographed Rigel rising in 2009 the sky was dark blue (Fig. 8.29). As with the northerly moonrise, the star's track projected backwards meets the true horizon at the calculated point (Fig. 8.30); but once again, the calculated track is too steep. On the actual horizon, the star would come up to the right of where I predicted. Once again, as with the variations in refraction, it was a mistake the Neolithic builders could never have made because they were recording in stone what they actually *saw*, averaged over generations of observation.

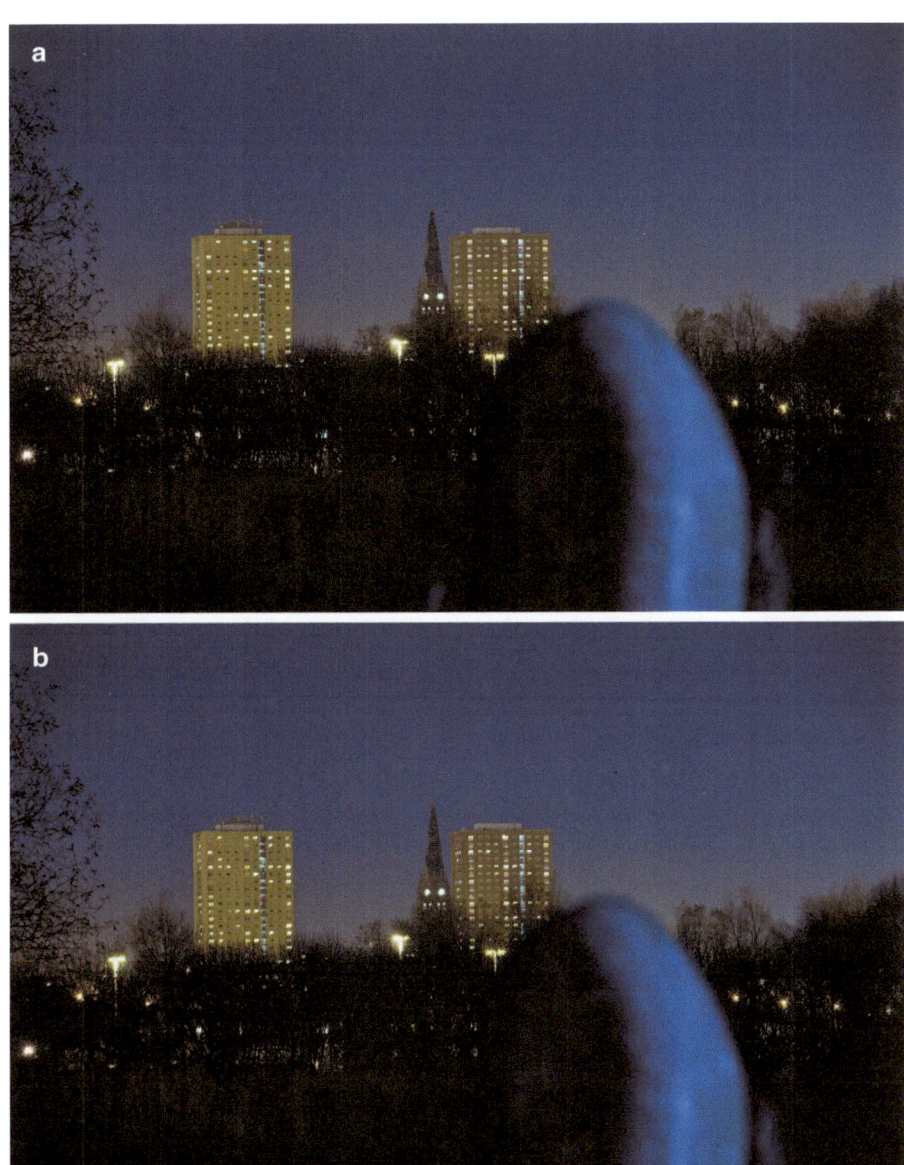

FIG. **8.29** (**a**, **b**) Rigel rising towards *upper right* (Photos by Mark Runnacles, November 30, 2009)

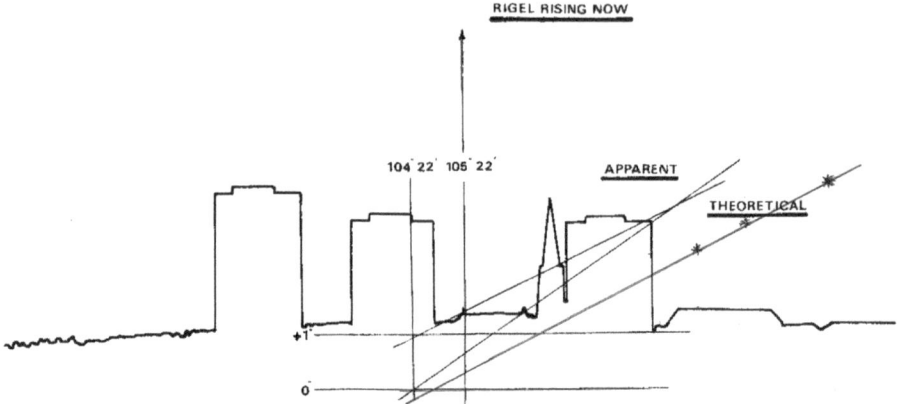

Fig. 8.30 Actual track of Rigel rising, 2009 (Drawing by Richard Robertson from working photograph, sketch by author)

What we learned from the solar events, and the more salutary lessons from the northerly Moon and Rigel, was an unexpected bonus. As stated above, Sighthill's intended contribution to the megalithic debate was rather different. The Sighthill megalith is intended primarily as a tribute to the ancient builders, and to the scientists who uncovered the significance of the work. It's less accurately aligned than the ancient sites for the practical reasons given above, and in twentieth century Glasgow its astronomical functions can in that sense be described as "ritualistic." Even after some hypothetical future collapse of civilization, Sighthill could not be used to construct an accurate calendar, unless perhaps by painstaking measurements of the length of the shadows.

Nevertheless it poses a challenge to critics of astroarchae-ology. In designing and building the Sighthill circle, with all the help and advice mentioned above, we strove to be faithful to the principles of the ancient sites as we understood them. There is no intended feature of Sighthill that does not have its counterpart or analogy in some ancient site. Yet we are here to testify that the design of Sighthill was primarily astronomical throughout – to incorporate solar, lunar and stellar alignments to the greatest accuracy attainable. Anyone who maintained that such was not the case with the ancient sites – that the astronomical alignments are non-existent, or coincidental – might reasonably have been asked to lay sociohistorical considerations aside and explain, in terms of the concrete evidence available, where the difference lay. With the change in perception of Neolithic society from the recent discoveries at the Bend of the Boyne, Brodgar, Kilmartin, Stonehenge and Durrington Walls, sociohistorical considerations now give that question much greater force.

References

1. Lloyd, A.L.: Folk Song in England. Panther Arts, 1969; Steeleye Span: Ten Man Mop, or Mr. Reservoir Butler Rides Again. Pegasus Records PEG 9 (1971)
2. Lunan, D.: Solar events at Sighthill. Griffith Observer 50(6), 2–11 & 20 (June 1986)
3. Lunan, D.: A stone circle for Glasgow. Tournaments Illuminated, 85, 32–35 (Winter 1987); updated, R.I.L.K.O. J. 78, 11–16 (May 2011), and reprinted online in The Stone Circle, www.sighthillstonecircle.net (June 2011)
4. Finch, B.: Talking with Tongues. Luath Press, Edinburgh (2003)

9 Archaeoastronomy from the Air

The hardest thing on earth to destroy evidence of is a hole in the ground.

— Leo Deuel, "Flights into Yesterday" [1]

"You'll be navigator, Duncan, OK?" said Leslie Banks as he taxied the Cessna towards the main runway of Glasgow Airport (Fig. 9.1). That was just the last stage in the escalation of my responsibilities, and somewhat startling, since my experience of light aircraft to date was limited to one 5-min hop around Turnhouse Airport in an RAF Chipmunk trainer, when I was in the Air Training Corps at 13 years old.

Ben Bova, the Science Editor of *Omni* magazine, had visited the stone circle in August 1979, while in the UK for the World Science Fiction Convention in Brighton. Since we had been told there was to be no further work on the circle, I was surprised to find the bank on the south perimeter was under construction, in accordance with the plans (Fig. 8.1). Bova commissioned an article from me, but after meetings with Dr. Bernard Dixon (*Omni*'s UK Editor, former Editor of *New Scientist*), and with staff reporter Kathleen McAuliffe, her article appeared instead in November 1981 [2].

It caught the attention of Leslie Banks, IBM's head of scientific public relations, whose hobby was aerial archaeology, and he offered to photograph the Sighthill circle for me from the air. He also invited me to IBM's Science and the Unexpected conference at Heathrow in March 1982. This was the first of nine extraordinary events, bringing together galaxies of top scientists, UK and world figures, to review the cutting edges and the controversies in a wide range of scientific fields.

Without looking at my files, names that come to mind include Sir Hermann Bondi, Prof. Jacob Bronowski, Prof. Richard

D. Lunan, *The Stones and the Stars: Building Scotland's Newest Megalith*, Astronomers' Universe, DOI 10.1007/978-1-4614-5354-3_9, © Springer Science+Business Media New York 2013

FIG. 9.1 The Cessna used in the aerial archaeology flight, hired by Leslie Banks (*center*) for photography by Chris Stanley (*right*) (Photo by author, July 28, 1982)

Dawkins, Dr. Richard Garwin, Prof. Thomas Gold, Prof. Stephen Jay Gould, Prof. Richard Gregory, Sir Fred Hoyle, Dr. Garry Hunt, Dr. Sergei Kapitza of the USSR, Prof. Eric Laithwaite, Prof. James Lovelock, and His Holiness the Dalai Lama. Archie Roy spoke on 'The Lamps of Atlantis' in 1986 (he's wearing his speaker's tie in Fig. 5.1), and I was the only amateur scientist to address a Heathrow Conference, on the Fermi Paradox in 1987 [3].

During the conference, Leslie expanded his offer, to cover other targets in the area which the astronomers involved in archaeoastronomy might request. When contacted about the flight. Roy, the Thoms and MacKie each had a particular request to make. For my part, as well as the Sighthill circle I was keen to include the standing stones on Colonsay, and also Temple Wood, Kintraw (Chap. 4) and Brainport Bay (see below), all of which I had visited on the tour organized by the Astronomical Society of Glasgow in 1978.

Meantime, however, the investigation into the Knappers structure at Clydebank had put me in touch with Graham Ritchie of the Royal Commission on the Ancient and Historical Monuments of Scotland, who traced the records of the project and lectured on them to the Archaeological Society of Glasgow in December 1981. Although he was skeptical about Mann's findings and archaeoastronomy in general, it turned out that we had another enquiry in common. While I had been investigating the standing stones on Colonsay as a check on Thom's theories (Chap. 5), Dr. Ritchie was also making a survey of them for more general research purposes.

I therefore passed on Leslie Banks' offer to Dr. Ritchie as well. As the Thatcher government's cuts had taken hold, the Ancient Monuments Commission no longer had any budget for excavations, much less for aerial photography, so he took our tentative flight plan around to every department inviting requests. The result was rather daunting: 18 targets at 12 locations! On top of all that we also had three target requests from the astronomy section of the Ayrshire Branch of ASTRA. It was starting to seem that the flight could make a real contribution to the field, and Leslie Banks agreed to give the requests priority. I was to handle the liaison, obtaining plans of the target sites and drawing up the suggested flight plan. But not knowing that I would be navigating the aircraft, I hadn't brought my file with me, nor the relevant Ordnance Survey maps.

July 28, 1982, started off bright and sunny, but there was haze moving across the country from the east coast. Driving from Wishaw to Glasgow Airport I came back into sunshine, but by the time the Cessna was in the air the airport and the city were overcast. We were not allowed to fly below 1,500 ft over the city, and the photographer, Chris Stanley, winner of the 1981 Vinten Prize for aerial photography, doubted whether the Sighthill photos would come out. He couldn't get a reading on his light meter at all. I had a Kodak Bantam Colorsnap Mark 1, which was beginning to show its age; the pictures taken under the overcast came out reasonably well (Fig. 9.2), but Chris's close-ups were impressive (Fig. 9.3), and for the rest of the trip unsurprisingly he had the edge.

I had made an appointment with the Parks Department management to explain about the flight and urge that the stone circle should now be landscaped into Sighthill Park. The site had been left to its own devices for 3 years and was now grassed over, but the bank on the south arc of the circle had never been stabilized and had now washed away downhill, so the stones on that side were

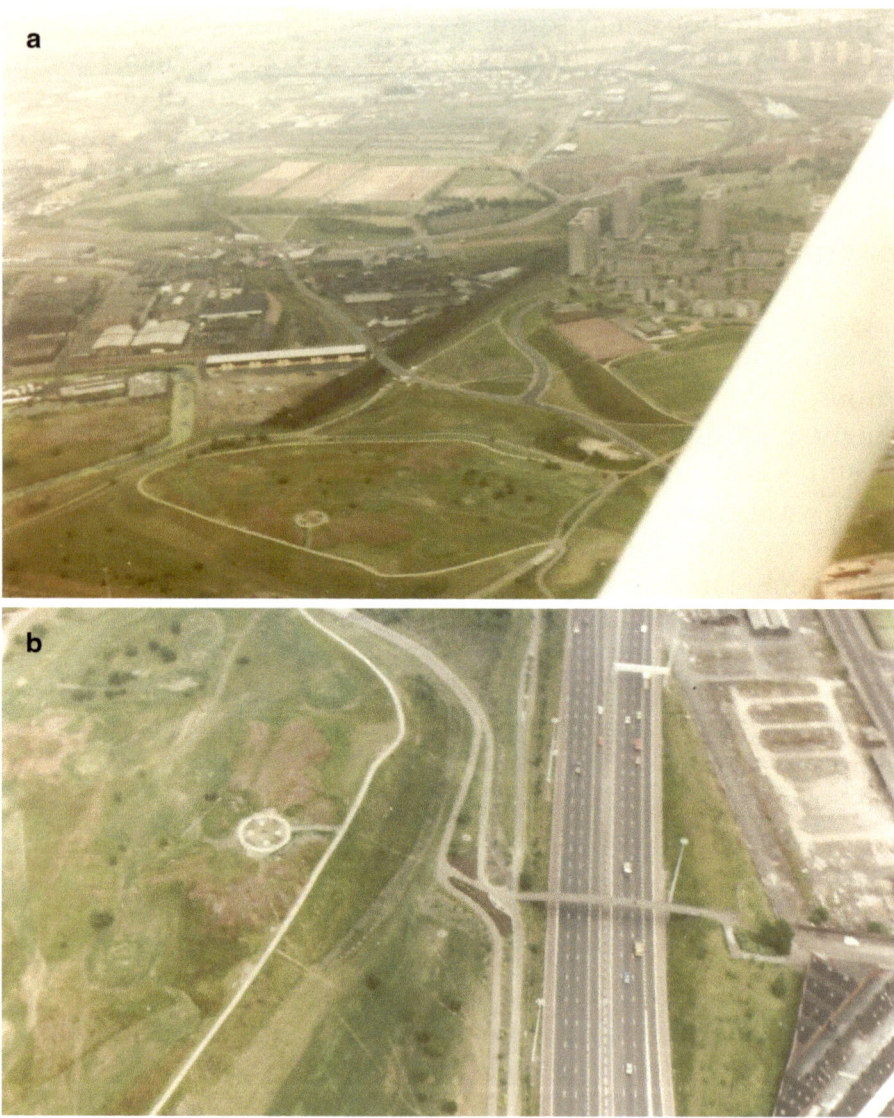

Fig. 9.2 (a) Sighthill Park, looking northeast. (b) The stone circle, east at top of picture, M8 motorway and footbridge to right (Photos by author)

once again too high to see over (Fig. 8.5b). Richard Robertson's drawing of the proposed 'cossie sets' layout was produced (Fig. 8.1), and it was agreed that to reduce costs, only the outer ring would be paved, plus the four decorative arcs, leaving the rest under grass. The go-ahead was given, but I wasn't there to supervise, and the

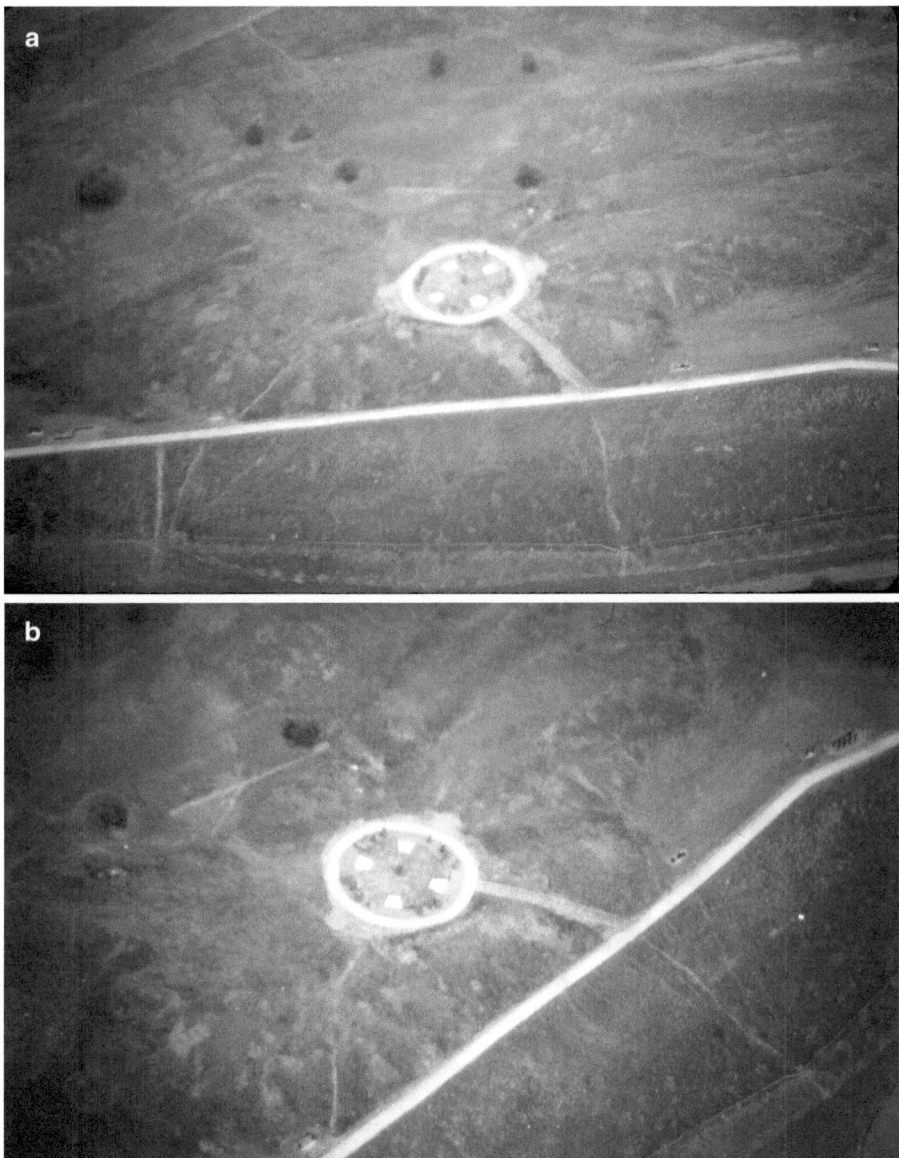

FIG. 9.3 (a) The landscaped stone circle (*north at top*), with a staircase lead-ing up 'Dead Man's Gulch' from the path below. (b) The circle looking northeast, towards midsummer sunrise (Photos by Chris Stanley, 1982)

plans were misread. To our disappointment, instead of recreating the bank the workmen flattened the contour of the site, leaving the stones on that side half-buried and much less spectacular, though it did look great from the air.

FIG. 9.4 Brainport Bay, near Minard, Loch Fyne (Photo by Chris Stanley, 1982)

As we flew west down the Clyde, coming back into sunlight, 'targets of opportunity' for Chris included a sunken freighter and a nuclear submarine in a floating dry dock. I took some general shots of the view as we crossed the Argyllshire hills. The main thing to come out of them was that even before the overcast arrived, there was a lot more haze in the atmosphere than was visible to the eye.

Given my inexperience, it was a relief when we found the first target without difficulty. MacKie's request was for coverage of the shore site at Brainport Bay, near Minard, on Loch Fyne (Fig. 9.4), where there is a long alignment of modified natural boulders (Figs. 9.5 and 9.7), plus two sighting stones that were found with their wedging chips and had been re-erected in their sockets (Figs. 9.6 and 9.7b). The line looks up the loch to the rising of the Sun at the midsummer solstice.

Fig. 9.5 Part of the long alignment at Brainport Bay (Photo by Duncan Lunan, 1978)

There are cup-and-ring markings on other nearby rocks, and off the main axis there's an elliptical cairn whose long axis seemed to mark midsummer sunset/midwinter sunrise, though it may in fact be relatively modern. Subsequent investigations found the site to have a major equinoctial alignment, but it has extra significance. From that position, the setting Sun would appear only once every 4 years, indicating that the site's creators had discovered the Leap Year. The original alignment would recur almost exactly after 33 years, and in 1976 a cache of 33 carefully selected white quartz pebbles was found hidden under a flat stone at the northeast end of the main alignment [4]. MacKie's critics accused him of grasping at straws to justify claiming an important astronomical use for the site; once again, readers may draw their own conclusions.

FIG. 9.6 (a, b) Markers for midsummer sunrise in the notch over Loch Fyne (Photos by author, 1978)

MacKie had asked for stereo coverage if possible, but that requires a special camera mounted under the wing of the aircraft and couldn't be arranged. My own snapshots were much better from here on – apparently it's easier to compensate for the motion of the aircraft when circling than when flying straight, as we had

FIG. 9.7 (a, b) Brainport Bay revisited. The contrast between light and shade, due to the increased tree cover, made the site difficult to photograph (Photos by author, 1997)

FIG. 9.8 (a, b) Netherlargie South Chambered Cairn, Kilmartin (Photos by Chris Stanley (1982) and author (1997)). (c) The stone circle at Temple Wood, Kilmartin, washed out by reflected sunlight (compare Fig. 8.2) (Photo by author, 1982)

been over the ships in the Firth – but there's still no comparison with Chris's pictures, taken with better equipment and with much more skill.

At Temple Wood and Kilmartin, we hit a different problem. The Sun angle was wrong, and the only picture Chris obtained was of the Chambered Cairn at Netherlargie South (Fig. 9.8). I photographed the nearby small circle, in which excavation had revealed a central kist, possibly the burial place of astronomer priests (Fig. 8.2); but the detail was washed out by the reflected glare from the stone bed (Fig. 9.8c).

Graham Ritchie had listed several other sites in Argyllshire, but as I wasn't familiar with them and time was passing, we made straight for Colonsay. Leslie let me take the controls in 'fly-by-wire' mode, on automatic pilot but overriding it, and I took us across the sea and through a preset descent on the approach to Balnahard Bay, on the northwest of the island. Over the brow of the hill to the south, Cnoc a' Charragh (the Hill of the Standing Stone)

FIG. 9.9 Standing stone and kist, Cnoc a' Charragh, Colonsay (Photo by Chris Stanley, 1982)

there's a very interesting stone with a slab beside it that covers an empty burial kist (Fig. 9.9). When I 'found' the site the Sun happened to be over a notch in the hills to the west (Fig. 9.10), shining on the sea and revealing that at that point there's a sea horizon. Rough bearings confirmed that it marked one of Thom's calendar dates (Chap. 5).

We next tried for the small circle overlooking the microwave dish of the telephone switching station at Scalisaig, and the possible sites of two more that I had found. I thought that photographing them from the air might reveal marks in the ground. But all of those stones are small ones, by megalithic standards, hard to see and briefly glimpsed from the air, even at 500 ft. I could see them because I knew precisely where to look, but Leslie and Chris could not.

Admittedly I became better at giving directions as the flight went on, but in planning such missions these are points to keep in mind. There was no difficulty in spotting the conspicuous hilltop stone at Garvard, on the south end of Colonsay, although in the photograph it's hard to distinguish from the sheep grazing around

Fɪɢ. **9.10** Notch to the west of Cnoc a' Charragh, with contrast enhanced (Photo by author, 1979)

it (Fig. 9.11a). I had found a line of stones leading north from it through the valley, leading to a cairn overlooking the abandoned village north of Scalisaig, so it was obviously used for local navigation. As well as the alignment to it, its long axis face now points to two prominent notches between the Paps of Jura (Fig. 9.11b). The direction is towards midwinter sunrise, and to moonrise at furthest south. Graham Ritchie told me that the stone fell in the 1930s and was re-erected in 1960. I suggested that there might be older photographs against which its present alignment could be checked, but I've heard nothing further.

At Kilchattan on Colonsay there's a real mystery – two very tall stones, perhaps 15 ft tall, placed crosswise in a valley so narrow that the hills, though low, rise high above them. The long axis

FIG. 9.11 (a, b) Standing stone, Cnoc Eibriggin, Garvard, Colonsay (Photos by Chris Stanley, 1982, and Julian Paren, May 1994)

FIG. 9.12 Mystery stones, Kilchattan, Colonsay (Photo by Chris Stanley, 1982)

of the valley points in one direction to a swamp, a loch and more hills, and to nothing striking in the other. Thom told me that he had visited the site but could make nothing of it. Ritchie has found seventeenth-century accounts of buried structures nearby, but we saw no marks with the naked eye or in the photos (Fig. 9.12).

The rising of the west of Britain after the Ice Age has left Colonsay with tiers of raised pebble beaches in the southwest, and Graham Ritchie had asked us to look for Viking grave sites. However, as we could see nothing and had many more sites to cover, we decided to press on. Each photo session involved circling for several passes, banking the aircraft and changing the under-carriage position to Chris's direction. Although the team became more efficient as the flight went on, taking pictures was a time-consuming business.

After we crossed the Strand, the ruins of Oronsay Priory gave Chris another target of opportunity that provided a spectacu-lar photograph (Fig. 9.13). Although my photograph of the 1940 stone on the strand made a point about its usefulness in misty conditions, my 'ground truth' photographs of the Priory in the rain

FIG. **9.13** Oronsay Priory from the air (Photo by Chris Stanley, 1982)

hardly do it justice (Fig. 9.14) South of the Priory Ritchie had asked us to photograph mysterious mounds, obviously of human origin. They look like burial sites, but nothing has been found within them. Nor, to be honest, does the photo seem to shed much light on the matter (Fig. 9.15).

On the approach to the island of Islay Ritchie had asked us to photograph a feature he had marked as 'Viking house?' by the ruined chapel on Nave Island. He hadn't given it enough priority even to describe it in his accompanying notes, but it proved to be the find of the mission. The remains of several former structures could be seen and the 'house?' itself is clearly boat-shaped.

The island's name may not relate to the chapel. The Vikings did build boat-shaped houses, but there may have been a ship burial somewhere in the area. Unusual numbers of brass rivets have been found in Viking graves on Oronsay, suggesting that a ship had been dismantled and the rivets used as currency. A key question about the age of the feature (which is less conspicuous at ground level) was whether it was crossed by the run-rig marks of Viking cultivation. If it was, the 'house' would be more recent. But in fact only one line crosses it, and that one is much coarser than the run-rigs (Fig. 9.16).

FIG. 9.14 (a, b, c, d) Oronsay Priory and Celtic Cross (Photos by author, 1978)

FIG. 9.15 'Burial mounds' south of Oronsay Priory (Photo by Chris Stanley, 1982)

FIG. **9.16** Ruined chapel and *boat-shaped* marking, not crossed by Viking run-rigs, Nave Island, off Islay (Photo by Chris Stanley, 1982)

As far as I know, what we seemed to have found has had no follow-up. The 'Viking House' near the chapel is listed as an ancient monument, but that seems to be all. Twice, when I've lectured on the aerial archaeology flight as a spin-off from the astronomy project, I've been approached by U. S. travel agencies interested in organizing digs. But they were thinking of 20–30 people, whereas Ritchie and MacKie both said that if they were to investigate it, they'd begin with at least a couple of seasons employing one researcher and a couple of assistants.

Ritchie had asked us to photograph his excavations of Viking graves at Bowmore on Ardnave Point, but at that point we had begun to have problems due to unfamiliarity with the sites. Luckily our photograph does show the trenches and vehicle tracks on the beach, and also an interesting polygonal feature in the grass, of which Ritchie was unaware (Fig. 9.17). But now we had a succession of disappointments as we failed to recognize the targets requested. In one case where we knew we had to be in the right area, we photographed a natural hill and missed the archaeological targets to either side of it. But we did succeed by chance

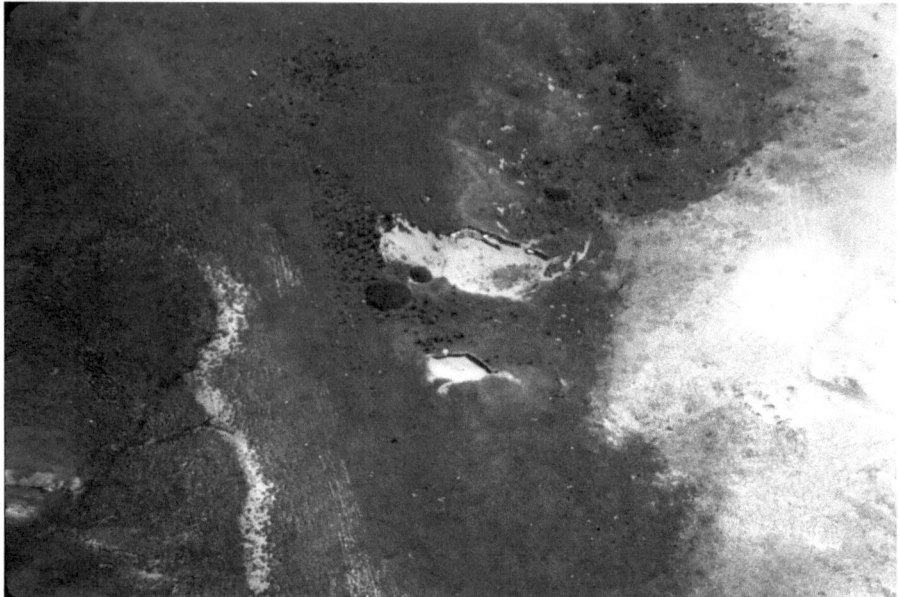

FIG. **9.17** Viking grave excavations on Ardnave Point, Islay (Photos by Chris Stanley, 1982)

in capturing a circular hill-fort on Islay that was another of his requested targets (Fig. 9.18).

I took the controls again between Islay and the Mull of Kintyre, and this time was allowed to cross the coast and make a turn, Leslie taking over when we asked permission to overfly Macrihanish Airport – which as he said must have surprised the radio operator on such a peaceful summer afternoon. But we couldn't find the standing stone to the south of it, and instead turned north to meet the Thoms' request at Pubal Burn, north of Campbeltown.

This time we showed that the camera can sometimes prove superior to the eye. In 1948 Archie Thom had seen the marks of a circle in the crops, and subsequently the stump of a stone was found. He had been trying to get an aerial photograph ever since! We saw nothing from the air, but when the slide was projected onto a large screen Christopher Taylor, a young member of ASTRA, found the faint marks of the former stones' positions (Fig. 9.19).

I passed on the photographs to Archie Thom, but the family had been struck by tragedy with the death of the grandson, Alasdair Thom, in a car accident, when the mantle of the family's

FIG. 9.18 Circular hill-fort by the telephone box (*at right center*), Islay (Photo by Chris Stanley, 1982)

FIG. 9.19 Field at Pubal Burn, Kintyre Peninsula. The marks of the former stone circle form an *ellipse at the center* of the field, but are barely visible in reproduction (Photo by Chris Stanley, 1982)

FIG. 9.20 Fingal's Cauldron Stones, Circle 5, Machrie Moor, Isle of Arran

concern with Neolithic astronomy was still settling about his shoulders. On my last visit to Dunlop with their copies of the aerial archaeology slides, Dr. Thom was friendly but inconsolable, and the Professor was imprisoned in the house he'd built himself, having broken his hip in a fall. It was a cold winter, and pneumonia, 'the old man's friend,' was waiting in the wings. It was the last time I was to see either of them in this world.

Roy's request was the megalithic complex at Tormore, on Arran in the Firth of Clyde. As stated in the quotation from his scientific thriller *Deadlight*, at the start of Chap. 1, the site has dozens of hut circles (stone walls with openings, once roofed with branches), but the main interest was the remains of astronomical structures. From a stone to the south the double ring of Fingal's Cauldron Stones – a perfect circle within a flattened one – lines up with the center of a flattened circle of the same shape to the northeast to mark the midsummer sunrise in a notch in the hills (Fig. 9.20).

All but one of the stones in the flattened circle have been reduced to stumps, but the survivor is 18 ft tall by 5 ft by 2,

FIG. 9.21 Surviving stone of the flattened circle, Circle 1, with smaller Circle 4 to the south (Photos by Chris Stanley, 1982)

comparable with the larger blocks at Stonehenge (Fig. 9.21). There's also a small ring to the south of it. East of the flattened ring, another large circle or egg has been even more badly damaged, and the remains of one stone still lie where it broke during an attempt to make it into a millstone (Fig. 9.22). A single stone to the west, the centers of the two large rings, and the southernmost of two small ellipses to the east, point to sunrise at the equinoxes. The ellipses, found by Roy himself and subsequently excavated by Aubrey Burl, were the first to be found in Britain (Fig. 9.23). About 26 are now known, including MacKie's excavation of the Cultoon ring on Islay (Chaps. 4 and 8), showing that the ancient Britons had discovered the ellipse 2,000 years before the Greeks around 600 B.C.

As mentioned above, ASTRA, the Association in Scotland in Research into Astronautics, was founded (originally as a Scottish branch of the British Interplanetary Society) in 1953 by the late Prof. Oscar Schwiglhofer from Transylvania. At that time Oscar's astronomical research included the history of Sir Thomas Makdougall Brisbane, who had an observatory in Largs, Ayrshire, in the

FIG. 9.22 (a, b) The three upright stones of Circle 2. Aerial photo by Chris Stanley, Vistamorph™ image by Chris O'Kane

nineteenth century, before his appointment as governor-general of New South Wales. An expedition organized by Brisbane discovered and named the river that in turn gave its name to the modern city. Brisbane's catalog of southern hemisphere stars is regarded as less valuable than Herschel's, but his reputation on the continent is higher than it is in the UK – hence Oscar's interest.

John Bonsor and Ron Williams of ASTRA's Ayrshire branch independently started research into Brisbane's activities in the 1980s, and had asked us to photograph the observatory site, as well as the "Three Sisters," three large meridian pillars on a hilltop in the town, with which Brisbane's three telescopes were lined

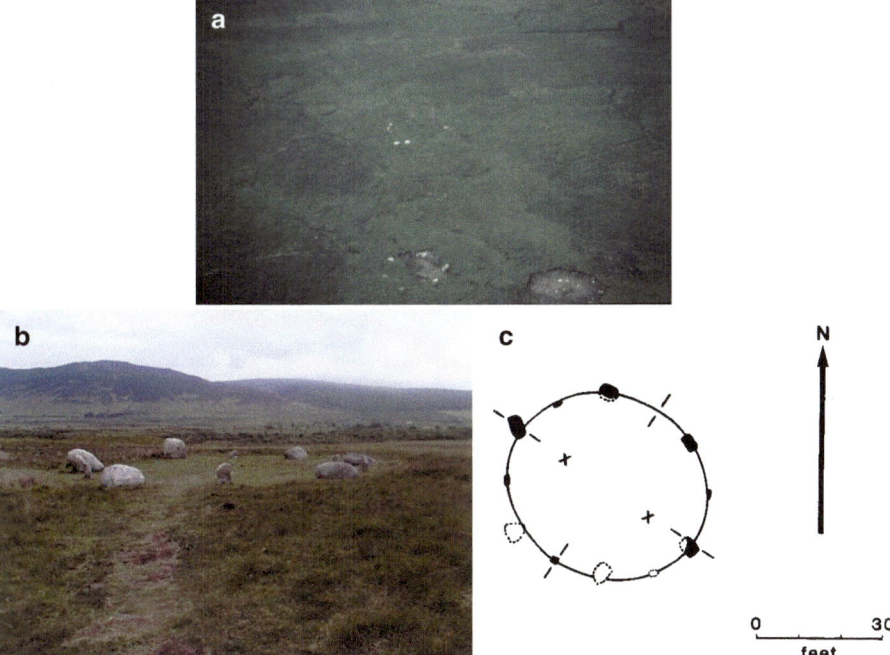

FIG. 9.23 (a) *Elliptical Circle* 1 (*left*), Circle 1A (*right*), Circle 2 remains above (Photo by Chris Stanley, 1982). (b) *Elliptical ring* at ground level (Photo by Chris O'Kane). (c) Tormore ellipse plan (Griffith Observatory drawing for 'Archaeoastronomy from the Air' [5])

up. Ron had also asked for coverage of a medieval village on the grounds of Hunterston House, adjacent to Hunterston nuclear power station. We thought we would be forbidden to overfly it, but did so unchallenged.

A last cautionary tale now emerges. I had passed Ron's maps to Leslie but somehow they had become separated from the flight package – and I did not have a copy. Duplicate everything! Although Chris was well pleased with the chance to photograph the reactor buildings and the nearby Hunterston Ore Terminal, once again we couldn't find the archaeological targets because we didn't know precisely where to look. We gained a good shot of the Three Sisters (Fig. 9.24), but I hadn't been to the observatory at the time, and since I knew it was overgrown but didn't know where it was there seemed little point in searching.

When I did visit the site, I pointed out to John Bonsor that the nearby walled garden of the estate (then under grass) would make

FIG. **9.24** The 'Three Sisters,' transit sighting pillars for the Brisbane Observatory, Largs, Ayrshire (Photo by Chris Stanley, 1982)

an excellent observatory site, screened from the lights of Largs by the highest part of the wall. In the 1960s I was involved in an observatory project, for a similar site in southern Ayrshire, which didn't come off. However an approach to the owner of Brisbane Mains Farm for permission to use the garden was kindly received, and in August 1983 the ASTRA Ayrshire Branch's observations of the Perseid meteors from the site were the most comprehensive in Scotland.

Lecturing to ASTRA in January 1984, Roy integrated the aerial photographs with his own shots of Tormore at ground level, to good effect. John Bonsor gave an entire lecture using only Fig. 9.24, projected on to a plan of the Brisbane Observatory and its transit sight-lines. The various researchers all requested copies of the Tormore slides in addition to their own targets, and Ritchie wanted copies of everything! Out of his 18 targets at 12 locations we had visited 8, drawing a blank at four, been frustrated by the light conditions at one, but obtained potentially useful photos at four as well as 'targets of opportunity' such as Oronsay Priory. We photographed Sighthill and three of my targets on Colonsay, met

MacKie's, Archie Thom's, and Roy's requests, and also covered the main objective at Largs. The flight was a fascinating experience, and what contribution the photos will make to the various inquiries remains to be seen (even now).

It's worth noting, however, that my involvement in the flight and before it in the Sighthill project, stemmed originally from the inquiry by my colleague Alan Evans, about the possible significance of markings on aerial photographs of Stonehenge. Aerial views of Stonehenge and Avebury can be found in virtually every book on archaeoastronomy, but few of the other megaliths seem to be covered. Aerial archaeoastronomy may well have a great deal to add.

References

1. Deuel, L.: Flights into Yesterday. MacDonald, London (1971)
2. McAuliffe, K.: Explorations: modern megalith. Omni 3(11), 118–120 (1981)
3. Lunan, D.: The Fermi paradox. Speculat Sci Technol, 11(1), 25–47 (Jan 1988); Updated, Asgard 4(4), 27–41 (Nov 2002)
4. MacKie, E.W. The prehistoric solar calendar: an out-of-fashion idea revisited with new evidence, op cit
5. Lunan, D.: Archaeoastronomy from the air. Griffith Observer, 50(11), 9–11 and 14–18 (Nov 1986)

10. The Circle, Present and Future

Your stone circle's worthless – it's *obsolete*.
                    – Local worthy, Townhead, Glasgow, 2010.

In the aftermath of the aerial archaeology flight, the big disappointment was the truncation of the stones, making the circle much less spectacular overall (compare Figs. 10.1 and 10.2). Before the cement dried, vandals also managed to scrape up some of it and smear it on the midsummer sunrise stone, where it remains to this day (Fig. 10.2a, background).

Salt was rubbed in the wounds soon after, by publication of a local guidebook saying, "The stones lack any real presence due to their lack of height." Other comments over the years have on the whole been more sympathetic [1]. I was told that the circle had acquired a nickname, 'The Cuddies' (horses), because now children could sit on them as if on a roundabout. Later I learned that it's a general name for the three hills, though I haven't found out how old the name is. If it dates back to the 100 Acre Hill Farm that's a little curious, since it was a dairy farm, but perhaps horses were reared there, too. However, stone rings very often take on the names of their locations.

Another form of laying claim to the circle is somewhat less welcome, though traditional in its own way. In the *Sunday Mail*'s color version of Ewen Bain's cartoon *Angus Og*, there was an episode where he was being paid by English tourists to show them the standing stones on the fictitious island of Drambeg. Seeing them silhouetted against the sunset, the tweedy woman cried, "Magnificent! Didn't Landseer paint them on his tour of the Utter Hebrides?" As the view shifted to the sunlit side of the stones, covered in graffiti, Angus replied, "No, it was some hooligans from Lewis at the shinty cup final."

D. Lunan, *The Stones and the Stars: Building Scotland's Newest Megalith,*
Astronomers' Universe, DOI 10.1007/978-1-4614-5354-3_10,
© Springer Science+Business Media New York 2013

Fɪɢ. **10.1** Author at the midsummer sunrise stone, originally one of the lowest in the circle. The midwinter sunrise arc, downhill on the far side of the circle, was then taller than head height (Photo by John Gilmour, summer 1979)

The Sighthill circle quickly acquired similar decoration. After accepting my article on "Solar Events at Sighthill" for the *Griffith Observer*, Dr. Ed Krupp visited Sighthill Park and took his own photographs to accompany it, with captions "Symbolic pictographs or undecipherable hieroglyphics have been added by an intrusive barbarian people.... The stones themselves display further evidence of invasion by a later, intractable tribe of philistines [2]." At first we were advised by the Parks Department to investigate cleaning and adding protective coatings for the stones.

After 10 years of the Clean Air Act, enacted to stop the killer smogs of the 1960s, Glasgow was beginning to shed the grime of the Industrial Revolution and centuries of coal fires. As the natural colors of the stone buildings began to show, their beauty became apparent. Sir John Betjeman had not been insane to call it the most beautiful Edwardian city in Britain. Going about the city during the project, John Braithwaite and I were repeatedly taken aback by the splendor of buildings we had taken for granted when they were all a uniform black.

The sometimes destructive fad for active stone cleaning did not sweep the city until a decade later, in the context of Glasgow's

FIG. 10.2 (a) Author at the half-buried midwinter sunrise arc, summer 1989 (Photo by Brian Fair). (b) Author at the midwinter sunrise stone, June 7, 2010 (Photo by Linda Lunan)

year as European City of Culture. The specialist companies that did exist in 1979 never replied to our queries. Eventually we consulted the Ancient Monuments Commission, who advised us not to do it but to let the stones weather naturally. Graffiti would wear off, but cleaning would destroy the natural patina. As it turned out, the stones are so exposed on the hilltop that each winter's storms scoured off the paint, including the occasional offensive messages, though new ones quickly replaced them.

I've occasionally been asked, by New Age believers not living locally, whether the presence of the circle has brought peace and prosperity to the area. During the project I was sent a copy of *Needles of Stone* by Tom Graves, who "demonstrates how he and his fellow dowsers can detect the natural energy patterns and flows associated with megalithic structures, and shows the connections between these patterns, ley lines, and traditional beliefs in ghosts, magicians, nature spirits and demons. Drawing parallels between his findings and Chinese acupuncture and geomancy – systematic alteration of the landscape – he shows how megalithic man could have used, or been used by, this knowledge to control the weather and improve the quality of life [3]."

You can't prove that by the standing stones on Colonsay. Three I haven't included were used imaginatively in more recent history, one by a firing squad, another for a pillory and a third for a whipping post. Nor has a beneficial effect on Sighthill been easy to detect (see below). But three visits to the circle by believers do stick in the mind.

The first assured me with great certainty that the Broomhill was a fairy hill – which might have been believed, at one time, though all that lies below it now is the disused railway tunnel. At one of the Glasgow science fiction conventions, I was asked to take a pagan group to the circle. They offered to connect it to the ley-line equivalent of the National Grid, which I politely declined, but going around it with them I was surprised to find the stones warm to the touch and giving what seemed almost like an electric charge. I'm not usually suggestible to that extent and went back 2 days later to check, but have felt nothing like it before or since. Tony Crerar was also a believer, as well as a serious observer, but related to a different set of traditional beliefs. Walking into the circle and observing the graffiti, he remarked, "Oh yes, this is building up its charge nicely. I bet that Brenda over there is pregnant by now, and I expect Mike over here had a lot to do with it..."

Originally Sighthill Park had been laid out with activity areas, linked by paths designated for runners, walkers and other fitness

FIG. 10.3 Forbes family memorial at the central stone, winter solstice 2009 (Photo by author)

devotees – featured originally on the plan which Dave McClymont modified for the helicopter operation brochure (Fig. 5.14). But with little take-up, the Park was instead allowed to develop into a nature reserve, with a wide variety of trees and plants, sheltering deer, foxes, ducks and other wildlife. Despite its original name of Broomhill, it hasn't been reclaimed by gorse. But even today, beautiful as it is, the park is strangely deserted even in summer when the other parks are crowded with people. The local family of the late Lily Forbes have been leaving memorials at the circle since her death in 2005, because it was one of her favorite spots, and they remain there undisturbed (Fig. 10.3).

So matters remained until 2000, when the Community Artist, Jim Campbell, decided to feature the circle in a mural for the Sighthill Youth Center (Fig. 10.4), and asked me to cut the tape at its unveiling (Fig. 10.5). After discussing the history of the circle with me, he suggested that we form a project to complete and renovate the circle. Our main aims were to restore the stones to their

Fig. 10.4 (a) Study by Jim Campbell for the Sighthill Youth Center mural. (b) Sighthill Youth Center Mural (Painting and photos by Jim Campbell)

intended height, with the tops at adult eye level, and to install a plaque explaining the circle. There was and still is nothing on the site to identify the circle, saying who built it, to whom it's dedicated or how it works, and after the initial affection of the 'Cuddies' nickname, by the twenty-first century the helicopter operation had been forgotten and local legend attributed it to the Druids. Supposedly local children were now afraid of it.

Originally we had planned to insert plaques into the surrounding paved walkway, with arrows pointing along the alignments and wording, 'Stand here for midsummer sunrise,' etc. Gavin Roberts had designed them and molds for the castings had been prepared, but the work was not allowed to be completed. A later plan, put forward by John Braithwaite and myself, was to use two of the unused stones as outliers to mark sunrise and sunset at the equinoxes, and use the two others to support a plaque bearing Dave McClymont's plan (Fig. 6.23), with a brief explanation as above.

FIG. **10.5** Mural unveiling, July 2000 (Courtesy of the *Springburn Herald*)

Jim Campbell and I resurrected that plan, and it found approval first with the Glasgow North Regeneration Agency, and then with the City Council, who added a requirement for better access to the circle – specifically, renewing the staircase leading to the circle and creating wheelchair access for the first time (Fig. 10.6). Highly constructive meetings were held with Land and Environment Services (formerly the Parks Department).

In 1999, to my dismay, the stones had been cleaned of graffiti and painted white, possibly another example of the 'City of Architecture and Design,' but destroying the first 20 years of natural patina they had accumulated. By 2001 the weather had again cleaned them off, and I secured a promise that there would no repetition. Plans were drawn up for the restoration, and the budget was agreed, with just two conditions. Jim and I had to obtain written support from community organizations in the area, and we had to obtain separate funding for community events involving the circle once it had been restored.

Almost immediately, the answer to both was provided. In response to rising tensions in the area, Strathclyde Police organized a multicultural festival centered on the Youth Center, where

SKETCH OF SITE PLAN

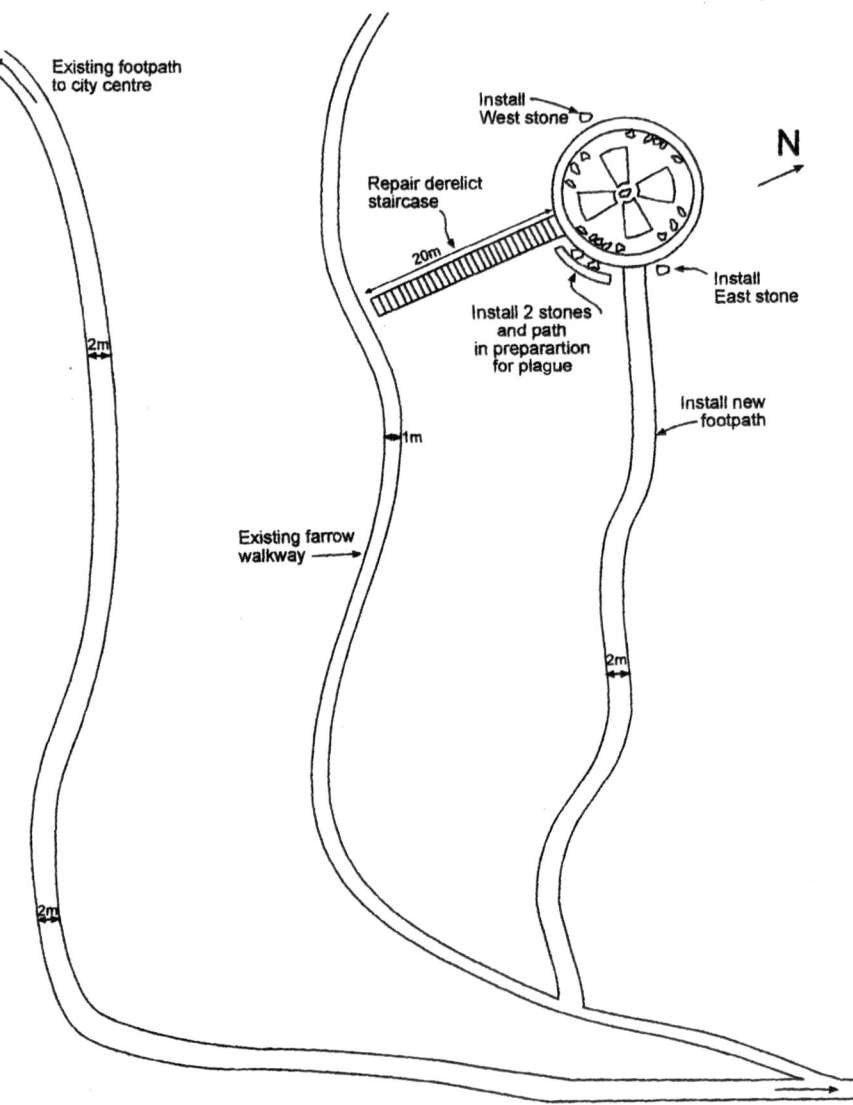

FIG. 10.6 Circle improvements plan. Land and Environment Services, 2001

the mural featuring the circle was so prominent. Jim Campbell and I took a table at the event and spent the day making contacts to obtain promises of local support. The Chinese community was especially interested, and I had some engrossing conversations about the origins of astronomy and the Great Wall (see Chap. 3). Around the same time, I was invited to lecture on the stone circle using the Zeiss projector at the Scottish Power Planetarium in the Glasgow Science Centre (see Chap. 2), and that drew more offers of support from community organizations in Sighthill. But then a major political setback arose.

The city of Glasgow was about to undertake a major new initiative to resolve its social housing problems. The mounting financial burdens of the outlying estates and the multi-story developments that had been created in the 1960s, in a misguided attempt to resolve the overcrowding and deprivation of previous schemes like the Gorbals tenements, had brought about a crisis in which (for example) household repairs had been frozen for 2 years because the budget for them had been consumed by interest charges. Glasgow's entire stock of social housing was shortly to be turned over to a new, nonprofit Glasgow Housing Association with an entirely different financial structure. But there was a way to get a head start on the problem of the high-rise flats.

Glasgow has a long tradition of absorbing refugees – for instance, at the time of the potato famine, and from Europe, particularly from Jewish communities, before, during and after the Second World War. None of those absorptions has been trouble-free. But during the Astronomy Project, and during a wave of anti-immigration riots in England, the late George Hay came up to lecture at the High Frontier exhibition and said to me, in the queue for a taxi at Central Station, "There's nothing you can show me that will convince me you don't have the same problems here." I walked across to a newsstand and came back holding open that night's *Evening Times* at the front-page headline "SCOTLAND WELCOMES THE BOAT PEOPLE."

Due to the conflict in the Balkans, in 2002 Britain was experiencing a new wave of asylum seekers, as refugees were now pejoratively termed. The government was offering large financial incentives to cities willing to take them, temporarily, and it created a major opportunity to improve Glasgow's housing stock. At its peak, about 6,000 refugees and asylum seekers were housed in flats regenerated from central funds; but, inevitably, those were in areas of deprivation, particularly Sighthill. It created great resentment among local residents, who had been told for years that they couldn't have emergency repairs, let alone new furniture or electrical appliances. Although the Strathclyde Police multicultural festival was successful and popular, tensions continued to mount, culminating in the stabbing of an asylum seeker from Eastern Europe on the footbridge over the M8, leading towards the stone circle, and his death on the footpath beyond.

Shortly afterwards, I was called to a meeting at the City Chambers with local Councilors for the Glasgow North wards, and in an eerie repetition of my turning down some of the sites offered in 1978 (Chap. 5), I was told that it would impossible to spend money publicly on the stone circle in Sighthill, when there were social problems of such magnitude to be addressed. It would be done, I was assured, but not until the crisis was over.

I had no opportunity to pursue the matter until 2008, when I was running the second phase of the North Lanarkshire Astronomy Project, centered on Airdrie Public Observatory where I was a curator, but with a much wider outreach. Financed by the National Lottery, initially with two pilot projects under Awards for All and now full-scale by Heritage Lottery, my colleague Bob Graham and I, with part-time helpers, were running a program of around 700 events with more than 450 schools visits. Two exhibitions in northeast Glasgow with accompanying school visits had convinced us that there was a need for a similar Greater Glasgow Astronomy Project, and with encouragement from the City Council, Heritage Lottery and the Scottish Parliament, we planned it around the renovation of the stone circle. The grant applications were submitted in December 2008, and by the spring of 2009, it was clear that due to the worsening economic situation, in the UK and worldwide, funding would not be forthcoming.

In 2010 I remarried, and my new wife Linda set about to make things happen, particularly in regard to the stone circle renovation. I gave a public lecture at the Ogilvie Centre at St. Aloysius Church, followed by a group visit to the circle for midsummer sunset (Figs. 8.7 & 10.7), which aroused extensive public and media interest.

We soon formed a new nonprofit organization, the Friends of the Sighthill Stone Circle, to bring the circle's restoration about, while Land and Environment Services gave the site a fresh clean-up (Fig. 10.8). We have begun a campaign to photograph sunrise and sunset at the equinoxes (Fig. 10.9), in hopes of erecting the marker

FIG. 10.7 The circle of people, summer solstice 2010 (Photo by Linda Lunan)

FIG. 10.8 (a) The circle cleaned, looking west towards Glasgow University (Photo by Mark McLaughlin, Land and Environment Services, October 1, 2010). (b) The circle cleared, looking northwest towards Port Dundas. (c) Midsummer sunrise arc, October 1, 2010 (Photos by Mark McLaughlin)

FIG. 10.9 (a, b) Equinoctial sunset behind the spire of Glasgow University (Photos by Linda Lunan, March 21, 2011)

stones that have lain unused under a tree near the circle since 1979 (Fig. 10.10).

Perhaps it's a good omen (recalling the demolition of the Pinkston Power Station's cooling tower) that in the last 2 years some of the Sighthill high flats near the circle have been demolished, clearing the northerly major standstill moonrise (Fig. 10.11) and the incinerator and the distillery on the Summerhill have gone likewise, clearing the midsummer sunset arc with its major and minor standstill moonsets (Fig. 10.12). The incinerator's fumes made their presence felt anywhere within a mile, whichever way the wind was blowing, and as soon as it shut down, the stones began to acquire a covering of lichen for the first time (Fig. 10.13).

Fɪɢ. **10.10** Unused stones lying in Sighthill Park (Photo by Linda Lunan, June 2011)

After the circle is restored and completed, what then, I am often asked. Back in 1979, our next plan was to build a model of the Solar System, scaled to the city of Glasgow. Gavin Roberts had realized that if the stone circle represented the Sun, then the orbit of Pluto (then still designated a planet) would be at the city boundary (Fig. 10.14). The orbits of Mercury, Venus, Earth, Mars, and a small arc of the orbit of Ceres, would all lie within Sighthill Park. Jupiter would be on the campus of the University of Strathclyde. Through the late Professor Tedford, of the Department of Electrical Engineering and of the Glasgow Astronomical Society, we had arranged for it to be placed next to the Steelhenge (Fig. 5.5), making a link between that and our circle.

Saturn would have been on Queen's Dock, now the site of the Scottish Exhibition and Conference Centre, but now obviously it would be on the other side of the river Clyde at the Glasgow Science Centre. Uranus and Neptune might be on Maryhill Road,

FIG. 10.11 (a) Sunrise at the circle, looking nor-nor-east, May 28, 2006 (Photo by Mark Runnacles). (b) Looking nor-nor-east, June 10, 2012 (Photo by Linda Lunan)

Fig. 10.12 Summerhill skyline after demolitions (Photo by Linda Lunan, June 10, 2012)

Fig. 10.13 Lichen growing on the northwest stones (Photo by Kate Braithwaite, autumn 2010)

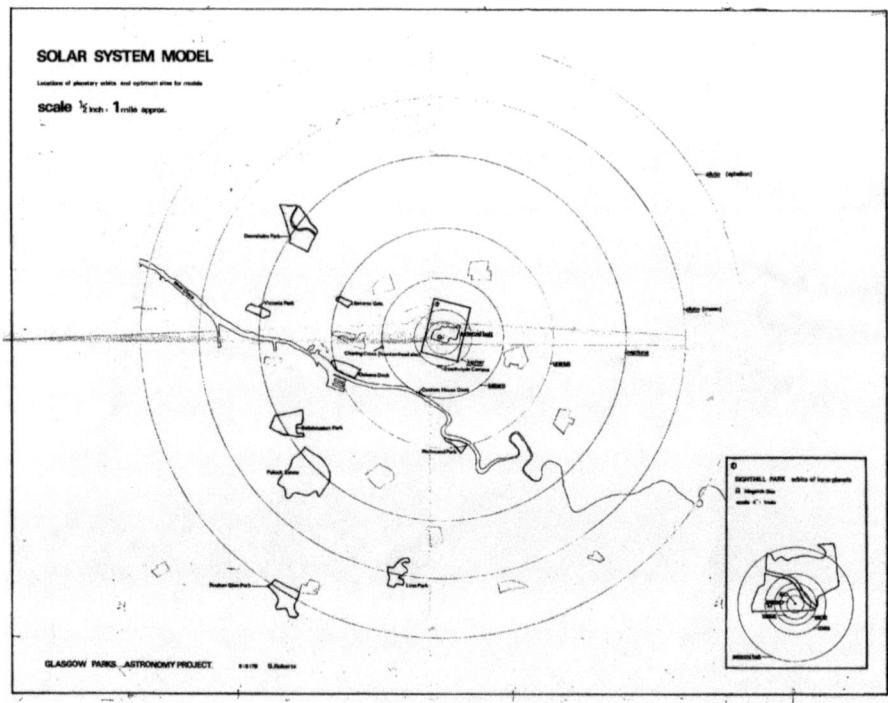

FIG. 10.14 Plan for Solar System model scaled to the city of Glasgow (Drawing by Gavin Roberts, 1979)

leading towards Glasgow University Observatory, and we thought there might have been a *Voyager 2* marker, moved progressively towards and past them over the next 10 years to 1989.

Pluto would be on Cathkin Braes, where my father mastered the game of golf in his youth. We thought it would be fun to have two more circles elsewhere in the world, perhaps in a municipal park in Vladivostok, marked 'Alpha Centauri, property of Glasgow Parks Department.' But it's a sobering thought that even on that scale, representing the Sun by a circle 40 ft across, the nearest star would be 40,000 miles away, off Earth's surface altogether.

Nevertheless, if the Sighthill stone circle is renovated and completed, the Astronomy and Space trail beckons next. If it were to happen in Glasgow, it could be extended across the Central Belt and into the Borders. There would be many visitor attractions to include. And extending it the other way, into the highlands and islands, perhaps it could provide the incentive to cherish and preserve the ancient sites – which is what Alexander Thom thought we should have been doing all along.

References

1. Moore, E.: Scotland: 1000 Things You Need to Know. Atlantic Books, London (May 2010); Holder, G.: The Guide to Mysterious Glasgow, op cit
2. Lunan, D.: Solar events at Sighthill, op cit
3. Graves, T.: Needles of Stone. Turnstone Books, London (1978)

John Braithwaite, 1945–2012, *iam exitu terre*. Photo by Jared Earle, 2011.

D. Lunan, *The Stones and the Stars: Building Scotland's Newest Megalith*,   309
Astronomers' Universe, DOI 10.1007/978-1-4614-5354-3,
© Springer Science+Business Media New York 2013

Glossary

Alignment Two or more objects marking the sight-line to an event such as the rising of a celestial body.

Altitude The distance of a heavenly body above or below the observer's horizon, measured in degrees and minutes.

Altazimuth Coordinates The position of a heavenly body in altitude and azimuth.

Amesbury Archer A.k.a. 'the King of Stonehenge': a man aged 35–45, buried c. 2300 B. C. 3 miles from Stonehenge, with archer's equipment, gold earrings and other marks of high status. He came from central Europe, probably Switzerland, and was missing a kneecap, probably from before he came to England.

Arran See Machrie Moor.

Asterism A grouping of stars, often in a distinctive shape such as the Summer Triangle or 'The Teapot' in Sagittarius. All constellations are asterisms, but not all asterisms are constellations.

ASTRA Scotland's national spaceflight society, founded in 1953 by the late Prof. Oscar Schwiglhofer as a Scottish branch of the British Interplanetary Society, becoming independent in 1963 and The Association in Scotland to Research into Astronautics Limited since 1976.

Aubrey Holes Exactly 56 pits within the ditch and bank of Stonehenge I, discovered by the Elizabethan/Stuart diarist John Aubrey, thought by Gerald Hawkins and Sir Fred Hoyle to function as an eclipse predictor.

Avebury The largest English henge monument, by Silbury Hill and due north of Stonehenge, with an initially huge ditch and bank ringed on the inside by sarsen stones, enclosing two smaller stone circles, and broken by four entrances, two of which were ceremonial avenues flanked by standing stones.

Azimuth The bearing of a celestial body measured from true north, either up to 360° (going eastward) or up to 180° east and west.

Ballochroy   A site on the Kintyre peninsula with three standing stones, lined up with a cist 120 ft away and to foresights marking midsummer and midwinter sunset over Jura and Cara Island.

BBC   British Broadcasting Corporation.

Beltane   Celtic festival held on April 30/May 1, marking the beginning of summer.

Bluestones   Spotted dolerite stones brought from the Preseli/Prescelly Mountains of Wales to Stonehenge and other sites en route, erected at Stonehenge II and rearranged within Stonehenge III.

Brochs   Circular Iron Age stone towers, possibly with astronomically oriented entrances.

Callanish   Large megalithic structure on the island of Lewis in the Outer Hebrides, part of a much larger complex of sites.

Campsie Hills or Fells   The prominent range of hills overlooking Glasgow on the north.

Candlemas   Christian feast, February 2, commemorating the Presentation of Jesus at the Temple. Also a key date in the megalithic calendar as reconstructed by Alexander and Archie Thom (see Imbolc).

Cardinal Points   The true north, south, east and west points on the observer's horizon.

Castle Rigg   A large flattened circle with multiple astronomical alignments, near Keswick in the Lake District of England.

Cathkin Braes   Slopes overlooking Glasgow on the southeast.

Celestial Poles   The positions where Earth's axis meets the celestial sphere at 90°′ declination north and south, overhead at the terrestrial poles.

Celestial Sphere   The entire sky pictured as a sphere of infinite radius centered on the observer.

Celts   Ancient peoples of western and central Europe, originating in the Iron Age Hallstatt culture of Austria, driven by later migrations into Brittany, Cornwall, Wales, Ireland and Scotland.

Circumpolar   Having a declination greater than the latitude of the observer and consequently always in the sky, never rising nor setting.

Cist or Kist   A box-shaped stone burial chamber overlaid with a stone slab.

Close In Glasgow, a passageway on the ground floor of a tenement building leading from the street to the stairwell and the back court or green (garden).

Constellation An area of the sky with fixed boundaries established by the International Astronomical Union in 1930, incorporating at least one of the classical constellations or the more modern ones of the southern hemisphere.

Cossie Sets or Setts Small granite blocks set in cement to form a cobbled path.

Cup-and-Ring Marks Carvings found sometimes on standing stones but more often on nearby rocks in their natural settings, possibly representing stars.

Cursus Latin for 'racetrack.' A very large rectangular enclosure dating from Neolithic times, true purpose unknown.

Cylinder Seals Rollers used to stamp one's identifying marks on Mesopotamian clay tablets, decorated with artistic symbols evolving over a period of 4,000 years.

Declination The distance of a celestial body above or below the celestial equator, measured in degrees and minutes.

Druids Celtic priesthood, leaders of rebellions against the Romans; users but not builders of Neolithic sites dating from 2,000 or more years earlier.

Drumlins Low, steep-sided hills in river valleys, deposited or shaped by glacial erosion during the ice ages.

Durrington Walls Very large Neolithic earthwork henge within 2 miles of Stonehenge along the river Avon, with a processional way to the river, postholes for large numbers of wooden uprights, possibly supporting large buildings, and over a thousand wattle-and-daub houses, occupied for no more than 50 years; the scene of huge feasts with cattle brought from all over the British Isles, all contemporary with the building of Stonehenge I.

Ecliptic The plane of Earth's orbit around the Sun, projected on to the celestial sphere, and hence the path which the Sun appears to follow through the year.

Ecliptic Coordinates The position of a heavenly body measured in degrees or minutes away from the ecliptic (ecliptic latitude) and from the vernal equinox (ecliptic longitude).

Ecliptic Poles   The two points in the sky 90° away from the ecliptic in ecliptic latitude. The north ecliptic pole is in the constellation Draco, near Boötes.

Equatorial Coordinates   Terrestrial latitude and longitude, projected onto the celestial sphere. The position of a celestial body measured in declination and right ascension, q.v.

Equinoxes   'Equal nights.' The two dates in the year, around March 21 and September 21, when the Sun is overhead on the equator and day and night are equal in length all over the world.

Foresight   Confusingly, a distant feature marking the position of a horizon event such as the rising of a bright star; by analogy with the foresight at the front end of a rifle barrel.

Galactic Center   The center of the Milky Way galaxy, occupied by the radio source Sagittarius A*, a supermassive black hole with about 4 million times the mass of our Sun.

Galactic Coordinates   The position of a celestial body measured in degrees and minutes along, and above and below, the galactic equator as viewed from our current position (see below).

Galactic Equator   The center line of the Milky Way, passing through the galactic center, as viewed from our current position north of the central plane of the galaxy. So, terrestrial-galactic coordinates do not correspond exactly to the true bearings of objects measured from the galactic center, especially for 'nearby' objects; but galactic coordinates measured at those nearby objects would virtually coincide with those measured here.

Gegenschein (Counterglow)   An apparent concentration of the zodiacal light, q.v., at the point directly opposite to the Sun's position on the celestial sphere.

Geocentric Coordinates   Celestial coordinates corrected for horizontal parallax at Earth's surface, q.v.

Gleniffer Braes   Hills overlooking Glasgow on the southwest, over Paisley and Glasgow Airport.

Grooved Ware   Flat-bottomed Neolithic pottery decorated by winding a cord around the wet clay object before firing, found particularly at wooden henge sites.

Henge   A circular monument enclosed by a ditch and bank. Thom was most insistent that a henge should not have a central feature such as a standing stone, but other writers are less punctilious.

Histogram A graphic representation of a frequency distribution, with vertical bars of widths corresponding to the class widths of the variable, and heights corresponding to the class frequencies. In other words, the more instances are found at particular values, the higher the peak on the graph.

Imbolc Celtic festival, now St. Brighid's Day, midway between winter solstice and spring equinox, marking the beginning of spring on February 1/2 (see Candlemas).

Kintraw A Neolithic site at Loch Craignish in Argyllshire, with a large cairn and a tall standing stone, aligned with midwinter sunset between the peaks of Jura as viewed from a platform of stones on a hillside overlooking the site.

Kist See Cist.

Lammas The Celtic festival of the wheat harvest on August 1.

Le Grand Menhir Brisé The Great Broken Stone at Carnac in Brittany, claimed to be the central foresight for surrounding lunar observatories.

Lintel A horizontal stone or wooden beam bridging two uprights.

Machrie Moor An area of the island of Arran in the Firth of Clyde, intensively settled in Neolithic times and having small astronomically aligned stone rings, including the first elliptical ones found in the British Isles, as well as the remains of very large ones.

Maes Howe A large stone-built passage tomb in a 24-foot high mound near the Stenness ring in Orkney.

Martinmas The Christian feast of St. Martin, November 11. Also a key date in the megalithic calendar as reconstructed by Alexander and Archie Thom.

Megalith A great stone, or a structure composed of great stones.

Megalithic Yard A standard unit of length in the layout of megalithic sites, postulated by Alexander Thom.

Menhir A single standing stone.

Meridian Pillars Outlying columns erected at a distance from an observatory, in a north–south line with the piers of the telescopes, to assist in aligning equatorial or altazimuth mounts.

Mesolithic An intermediate period between the Paleolithic Old Stone Age and the Neolithic New Stone Age. The development of boats and fishing led to a major spread of Mesolithic culture around the coasts of the British Isles.

MoD   UK's Ministry of Defense.

Monolith   A single stone.

MSC   The Manpower Services Commission, part of the UK Department of Employment, set up in 1973 to promote employment-related education and training, done away with in 1987.

NALGO   The National and Local Government Officers' Association, trade union for Glasgow Parks Department employees in the late 1970s, when it was largest trade union in the UK with over 700,000 members. NALGO was one of three unions that merged to form UNISON in 1993.

Neanderthal   Form of humans whose last common ancestor with *homo sapiens* was c.1,000,000 years before present. Driven out of Britain by advancing ice around 360,000 years ago, contemporary with modern humans in Europe around 100,000 years ago, re-entered Britain around 80,000 years ago, coexisted with modern humans until they became extinct 36,000 to 24,000 years ago.

Neolithic   The New Stone Age, the period following the Paleolithic Old Stone Age and intermediate Mesolithic period, q.v., and preceding the Bronze Age.

Nodes   The two points where two great circles intersect on the celestial sphere. The nodes of the ecliptic and Earth's equatorial plane mark the positions of the Sun at the vernal and autumnal equinoxes; the ascending and descending nodes of the Moon's orbit mark the points where the Moon crosses the ecliptic going north or south, respectively. Eclipses can only occur when the new Moon or full Moon are near the lunar nodes.

Obliquity of the Ecliptic   Another name for the tilt of Earth's axis, relative to the perpendicular of the plane of its orbit around the Sun. The obliquity varies between 24° and 22° with a period of about 41,000 years and has decreased by approximately half a degree since Neolithic times.

Orrery   A model displaying the workings of the Solar System, named after the Irish earl who commissioned the first known example in 1704.

Parallax   The effect of perspective that changes the apparent position of an object, relative to more distant objects. On the celestial sphere the effect is greatest at the horizon (horizontal parallax), and although negligible for the Sun and stars, it markedly changes the apparent position of the rising or setting Moon, relative to the bearing measured from the center of Earth.

Planisphere A representation of slightly more than a hemisphere of the sky on a flat plate, often with a mask that can be rotated to reveal only the stars visible at particular dates and times. A planisphere shows the relationships of the stars accurately in declination, but increasingly distorted in right ascension towards the edges, like the polar regions on a Mercator projection map of the world.

Plough, the A.k.a. the Big Dipper or the Drinking Gourd, in the United States, but depicted as a plough on Mesopotamian cylinder seals; the seven brightest stars in Ursa Major, and part of an open cluster through which our Sun is currently passing.

Polaris The brightest star in Ursa Minor, the northern pole star of modern times, now at its closest to the celestial pole – see precession of the equinoxes.

Posthole A prepared socket for an upright standing stone or wooden pillar; or simply the hole left behind by an upright post.

Precession of the Equinoxes A 26,000-year wobble of Earth's axis, caused by the pull of the Sun and Moon upon Earth's equatorial bulge. Although the current north pole star is Polaris, in 3000 to 2500 B. C. it was Thuban in Draco, and 13,000 years ago it was Vega in Lyra.

Proper Motion The movement of a star across the celestial sphere, seen from here, as a component of its true motion in space. The stars in open clusters such as the Pleiades, Hyades and the Plough, q.v., have shared proper motion with individual components added.

Radiocarbon Dating A method of determining the age of organic material by the decay of the radioactive isotope carbon-14, formed in the upper atmosphere by cosmic rays and solar radiation. Discrepancies between C_{14} dates and known historic dates were largely removed in the late 1970s when it was realized that the ratio of carbon-14 to other isotopes in the atmosphere varies considerably in proportion to solar activity. Correlations with tree rings, ice cores and other indicators have pushed back the dates for Neolithic Britain to a range starting well before the unification of Egypt around 3100 B. C., much older than the Minoans and Mycenaens, who were once thought to have built or inspired the British megaliths.

Recumbent Circles Stone circles found particularly in northeastern Scotland and in Ireland, where the stones are graded in size away from a large, horizontal stone between two vertical uprights, possibly so that the southerly Moon can appear to roll along the top of the horizontal stone.

Regression of the Lunar Nodes   The westward motion of the nodes where the Moon's orbit crosses the ecliptic, as the orbital plane precesses in space due to the pull of the Sun and of Earth's equatorial bulge. Each complete circuit of the ecliptic by the nodes takes 18.61 years.

Right Ascension   The equivalent of terrestrial longitude, projected onto the sky, and measured in hours and minutes from the vernal equinox, q.v.

Samhain   Celtic harvest festival, October 31 to November 1.

Sarsens   Tertiary sandstone boulders brought from the Marlborough Downs, 20 miles north of Stonehenge, and used in the construction of the Stonehenge III lintels and trilithons; also for the stone circles at Avebury.

Silbury Hill   A conical artificial hill erected near the site of the Avebury complex around 2700 B. C. The construction technique is analogous to the one used for the stable Egyptian pyramids, 2650 to 2500 B. C. Excavations have found no interior chambers or burials within Silbury Hill.

Skara Brae   A Neolithic village of nine stone huts on the largest island of Orkney, built around 3200 B. C. and abandoned around 2400 B. C., buried by sand and uncovered by a storm in 1850.

Solstices   The dates in the year when the Sun reaches its most northerly and southerly declinations, around June 21 and December 21.

Standstill   One of the two occasions, in the 18.61-year cycle of the regression of the lunar nodes, when the Moon reverses its monthly shift in declination. At major standstill the Moon at its furthest north for the month rises to the north of the midsummer Sun, and at minor standstill 9.3 years later it rises to the south of it at the corresponding point in the month, both times with a corresponding southerly extreme 14 days later.

Temple Wood   A site in Kilmartin Glen, Argyllshire, with large standing stones that Alexander Thom believed to be the most sophisticated lunar observatory in the British Isles. There's also a large, but low, stone circle with a bed of selected stones and a central kist.

Trilithon   A free-standing structure of three stones, two upright and the third forming a lintel across them. The larger inner structure of Stonehenge III consisted of a horseshoe of trilithons on the midsummer sunrise/midwinter sunset axis.

Tropics   The band of latitudes on Earth within which the Sun passes overhead in the course of the year. In each epoch its extent, north and south, equals the value of Earth's axial tilt (the obliquity of Ecliptic) and the radius of the Arctic and Antarctic Circles. The Sun is overhead at the Tropic of Cancer on midsummer's day in the northern hemisphere, around June 21, and at the Tropic of Capricorn on northern midwinter's day (midsummer's day in the southern hemisphere), around December 21.

Vernal Equinox   The date at which the Sun crosses the celestial equator moving northward, in the direction of increasing right ascension, around March 21 each year; by extension, the position of the Sun on that date, currently located in the constellation Pisces but confusingly called 'the First Point of Aries' for historical reasons dating back to ancient Greece.

Whinstone   A variety of basalt quarried at Kilsyth and at Hillhouse Quarry, Troon.

Wiltshire Lozenge   A bronze plaque, thought to have been worn as a breastplate, found at Bush Barrow immediately south of Stonehenge in 1808; interpreted by Archie Thom as an astronomer's template, but allegedly flattened and polished in a 'restoration' that removed the crucial markings.

Zenith   The point on the celestial sphere directly above the observer, at an altitude of 90°.

Zodiac   The band of constellations, centered on the ecliptic, within which the Moon and the planets normally move; the only ones that can pass outside it are Mercury and Pluto (which is no longer officially a planet).

Zodiacal Light   A diffuse band of light from the Sun scattered around the plane of the ecliptic by interplanetary dust, linking up with the Gegenschein opposite to the Sun, from the observer's viewpoint. Zodiacal light is often depicted by artists as twin cones to either side of the Sun, but actually it is only visible before sunrise and after sunset, particularly in the tropics. Until the mid-1970s it wasn't known whether the effect was local to the Earth-Moon system, but the *Pioneer 10* space probe found that it persisted out to the orbit of Jupiter, showing the dust comes from the Asteroid Belt and from comets passing through the inner Solar System.

Index